SpringerBriefs in Computer Science

Series editors

Stan Zdonik
Peng Ning
Shashi Shekhar
Jonathan Katz
Xindong Wu
Lakhmi C. Jain
David Padua
Xuemin Shen
Borko Furht
V. S. Subrahmanian
Martial Hebert
Katsushi Ikeuchi
Bruno Siciliano

For further volumes:
http://www.springer.com/series/10028

SpringerBriefs in Computer Science

Mohammed M. Alani

Guide to OSI and TCP/IP Models

 Springer

Mohammed M. Alani
Middle East College
Muscat
Oman

ISSN 2191-5768   ISSN 2191-5776   (electronic)
ISBN 978-3-319-05151-2   ISBN 978-3-319-05152-9   (eBook)
DOI 10.1007/978-3-319-05152-9
Springer Cham Heidelberg New York Dordrecht London

Library of Congress Control Number: 2014932534

© The Author(s) 2014
This work is subject to copyright. All rights are reserved by the Publisher, whether the whole or part of
the material is concerned, specifically the rights of translation, reprinting, reuse of illustrations,
recitation, broadcasting, reproduction on microfilms or in any other physical way, and transmission or
information storage and retrieval, electronic adaptation, computer software, or by similar or dissimilar
methodology now known or hereafter developed. Exempted from this legal reservation are brief
excerpts in connection with reviews or scholarly analysis or material supplied specifically for the
purpose of being entered and executed on a computer system, for exclusive use by the purchaser of the
work. Duplication of this publication or parts thereof is permitted only under the provisions of
the Copyright Law of the Publisher's location, in its current version, and permission for use must
always be obtained from Springer. Permissions for use may be obtained through RightsLink at the
Copyright Clearance Center. Violations are liable to prosecution under the respective Copyright Law.
The use of general descriptive names, registered names, trademarks, service marks, etc. in this
publication does not imply, even in the absence of a specific statement, that such names are exempt
from the relevant protective laws and regulations and therefore free for general use.
While the advice and information in this book are believed to be true and accurate at the date of
publication, neither the authors nor the editors nor the publisher can accept any legal responsibility for
any errors or omissions that may be made. The publisher makes no warranty, express or implied, with
respect to the material contained herein.

Printed on acid-free paper

Springer is part of Springer Science+Business Media (www.springer.com)

Foreword

Standards are needed to promote interoperability among vendor equipment and to encourage economies of scale. Because of the complexity of the communications task, no single standard will suffice. Rather, the functions should be broken down into more manageable parts and organized as communications architecture. The architecture would then form the framework for standardization. This line of reasoning led the International Organization for Standardization (ISO) in 1977 to establish a subcommittee to develop such an architecture. The result was the Open Systems Interconnection (OSI) reference model. Although the essential elements of the model were in place quickly, the final ISO standard, ISO 7498, was not published until 1984. A technically compatible version was issued by CCITT (now ITU-T) as X.200.

Meanwhile, a somewhat different architectural model emerged from the development of Internet-based protocols, and goes by the name TCP/IP model. Because rapid progress was made in the development of protocols within the TCP/IP model and in the deployment of such protocols by numerous vendors, TCP/IP became established as the "practical" model for communications protocols. OSI, on the other hand, can be viewed as a "descriptive" model that provides a universal terminology and context for discussing protocol operation. Thus, TCP/IP and OSI are complementary.

In this monograph, Prof. Mohammed M. Alani provides an excellent introduction to both OSI and TCP/IP. Chapter 1 provides a context for the discussion, by introducing computer networking and the concept of a layered model. Chapter 2 examines the OSI model. This chapter looks in detail at the functionality of each of the layers and provides a clear example of how data travel through the layers from a source system, through intermediate systems, and to the destination system, showing the actions at each layer along the way. Finally, Chap. 3 looks in detail at the functionality of each of the layers of the TCP/IP model and describes a number of important protocols that have been implemented within this model.

All in all, Prof. Alani has provided a concise, well organized, and clear introduction to these important networking topics.

William Stallings

Preface

Computer networks have become an integral part of our daily life. As we rely on networks more, we need to make a better understanding of the underlying technologies that provide us with these services.

The concept of a layered model makes it much easier to study networks and understand their operation. The distinction and clear separation of functions for each layer also makes the process of designing protocols much easier. The logical separation of layers makes people's lives much easier when troubleshooting. It makes it sensible to be able to isolate the problem and troubleshoot it much faster.

ISO's OSI model has been around since the early 1980s. Although it did not succeed in becoming the de facto model of networking, it is considered an important concept that helps a great deal when it comes to understanding how networks operate. The concepts presented in the OSI model help anyone interested in starting a journey into the world of networking. Back in the 1980s, OSI was gaining momentum and seeing it as a worldwide standard seemed very imminent. However, as Andrew L. Russell puts it, "by the early 1990s, the (OSI Model) project had all but stalled in the face of a cheap and agile, if less comprehensive, alternative: the Internet's Transmission Control Protocol and Internet Protocol."

TCP/IP model came in as a simpler, less-restrictive, and cheaper alternative. After looking like the savior of the world of telecommunications, the OSI model started to seem too comprehensive and too restricting. The fast-paced developments in the world of electronics and communications demanded a parallel standard for network systems that are easier to work with and are less demanding. Most entities involved in the networking world starting from computer scientists and ending with industrial partners have shifted belief to the TCP/IP model.

This brief starts with a simple introduction to computer networks and general definitions of commonly used terms in networking in Chap. 1. The last part of the chapter discusses the reasons behind adopting a layered model.

Chapter 2 discusses in detail the OSI model starting from a brief history of the standard. The concept of connection-oriented and connectionless communications is also introduced in this chapter. Subsections of the chapter elaborate on the specific layer functions and what is expected of protocols operating at each layer. In the last part of the chapter a detailed step-by-step description of how a single packet travels from the source to the destination passing through a router is explained.

Chapter 3 is devoted to the TCP/IP model. A better understanding of the model lies in better understanding of the protocols constituting it. Thus, the chapter starts with a discussion of IP protocols and its supporting protocols: ARP, RARP, and InARP. This discussion explains the details of the IP packet and the operation of the IP protocol. The next section explains the two protocols operating at the transport layer: TCP and UDP. The details of each protocol segment are introduced and functions of each field in the headers are explained. The next section discusses the detailed inner working of application layer protocols like HTTP, DNS, FTP, TFTP, SMTP, POP3, and Telnet. Details on how each of these protocols operates are also introduced. The messages and server response types for each application layer protocol are discussed.

Intended Audience of the Brief

- Students starting study in the networking area.
- Professionals seeking knowledge about networking essentials.
- Field engineers working in troubleshooting on an application level.
- Researchers looking for core concepts of networking.
- Anyone interested in understanding how Internet protocols are used in everyday life work.

How to Use This Brief

If you are new to networking and need to build a solid theoretical knowledge of networking, you should start from Chap. 1 and follow on to the following two chapters. If you are looking for gaining knowledge about application protocols like HTTP, FTP, etc., jump directly to Chap. 3.

The brief contains small gray boxes that are meant to emphasize the important definitions or facts that are thought essential to the reader before going further in reading.

Finally, I would like to thank my editors Wayne Wheeler and Simon Rees. Without you guys this publication would not be possible. Thank you for believing in me.

I would like also to extend my thanks to Prof. William Stallings for taking the time to go through the manuscript and writing the Foreword.

My final thanks go to my family, Marwa, little Aya and Mustafa, Mom and Dad. Thank you all for enduring me during the time of working on this brief and all my life. I couldn't have been blessed more.

Muscat, Oman, January 15, 2014   Mohammed M. Alani

Contents

Chapter 1
What are Computer Networks?

Abstract This chapter starts by defining networks and their uses. The reasons behind using a computer network are also introduced in this chapter. This is followed by definitions of the most commonly used terms that a starter needs to know. The first section ends with an introduction to the three modes of communication; simplex, half-duplex and full-duplex. At the end of the chapter, the reasons behind studying networks as a layered model are identified.

Keywords Network · Full-duplex · Half-duplex · Simplex · Protocol · Layers

1.1 What is a Computer Network?

A computer network can be defined as a group of hosts connected together to accomplish a certain task. The host can be a computer, a network printer, a server, or any other device that can communicate within the network. To keep everything in order, this network has to be governed by one or more protocols. A protocol is a set of rules governing the communication between two or more hosts.

The big question that pops up is why we would need a computer network. The answer is simpler than you think; it is to share resources. These resources may vary from information displayed on a web page, to just an empty space on a server's hard drive, to a printer, to anything that is on one host that can be useful to someone else. So, it is all about the resources.

> A Protocol is a set of rules that govern the communication between two or more entities.

In order to operate properly, the computer network needs some supporting devices. Devices such as hubs, switches, and routers utilize the operation of a network to make accessing the services easy and consistent. Sometimes these

M. M. Alani, *Guide to OSI and TCP/IP Models*, SpringerBriefs in Computer Science, DOI: 10.1007/978-3-319-05152-9_1, © The Author(s) 2014

devices are also used to control the network operation to guarantee that the network services will be used in the right manner and by the right people. Some other times these devices are used to monitor the network operation to detect any unwanted activity.

Networks may vary in size from a small office network with four or five computers to millions of computers, such as the *Internet*. This variation of size gives a wide variation of the services introduced. You can connect to a network to copy a small daily report to your boss's computer and you can also connect to a network to have a video conference with a colleague in the other half of the world.

> A common misunderstanding is that the World-Wide Web (WWW) is the Internet. The WWW, or the webpages, is only one of tens of services provided by the Internet.

As networks develop rapidly, new services also evolve, but also new challenges arise. These challenges can be the lack of bandwidth, new security threats, or merely the need for a new more powerful hardware.

It would be out of sense if we talk about computer networks without mentioning the largest network in the world; the Internet. The Internet is basically a huge network that consists of a large number of smaller networks. It connects millions of hosts together. Many people think that the World Wide Web (WWW) is the Internet. Well, they are wrong. The WWW is only one service of the many services the Internet provides. Examples of these services are electronic mail, file transfer, voice transmission, and many more.

The following sections will introduce basic definitions of what you need to know before going further into the network models.

1.2 Definitions

Now we will go through few simple definitions of some terms that we will be using later:

- Bandwidth: The maximum possible rate of data transmitted over a channel. This rate is measured by bits per second (bps) and its multiples, Kilobits per second (kbps), Megabits per second (Mbps), and Gigabits per second (Gbps).
- Throughput: The actual rate of data transferred between two hosts in a network without errors. Throughput is measured in the same units of bandwidth. Throughput of a certain network can not be greater than the bandwidth of that network.

- Host: A device that can communicate with a network. This device can be a computer, a server, a printer, or any other device that has the capability to communicate with a network and has the required set of protocols.
- Internetwork: A relatively large network that is a product of connecting two or more smaller networks.
- Local Area Network (LAN): A network that connects a group of hosts within a limited geographic area.
- Wide Area Network (WAN): A network that connects hosts over a large geographic scope. This type of networks usually uses carriers to deliver data from one host in the network to the other.
- Network device: A device that supports the network operation and helps in transporting the data correctly from one host to another. Examples of network devices are repeaters, hubs, switches, and routers.
- Link: A physical connection between two or more devices. If the link is between two devices only, it is called a dedicated link or point-to-point link. And if the link is between more than two devices, it is called a shared link or a broadcast link.
- Network Medium: A physical medium connecting hosts and networking devices. The medium can also be dedicated between two or shared among more than two entities. Examples of the medium are Unshielded Twisted-Pair (UTP) cables, fiber cables, and even the air (and the void space) is considered the medium for the wireless networks.

1.3 Communication Modes

There are basically three modes of communication in all communication systems; simplex, half-duplex, and full-duplex (sometimes full-duplex is referred to as duplex). Figure 1.1 shows the three modes of communications.

Simplex communication involves the transmission of data in one direction all the time. An example is listening to a radio station. The data flows only from the station transmitter antenna into you radio device, but you can not send data in the opposite direction.

In *half-duplex* communication, the two parties share the same communication channel to send and receive data, but on time-sharing basis, i.e. when X sends data to Y, Y can not send data to X at the same time. Y will have to wait until X is done and the communication channel is free to send data to X. So, the data flows in one direction only at a certain time. An example of this mode of communication can be seen in walky-talkies, or two-way radios. Only one user can send data at a certain time, and the other user can send data when the channel becomes free.

Full-duplex communication uses two separate channels for transmission and reception at each end. This means that data can flow in both directions at the same time. The telephone is a clear example of full-duplex communication as voice signals pass in both directions at the same time.

Fig. 1.1 Communication
modes: **a** Simplex, **b** half-
duplex, **c** full-duplex

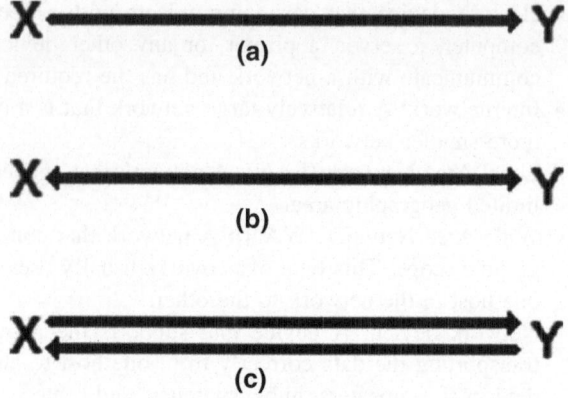

1.4 Why a Layered Model?

Computer networks are complicated, and they require harmony between different elements in order to operate smoothly. Some of these elements are hardware elements and others are software elements.

The network can be divided into parts to ease the understanding of its operation. In order to make these parts comprehendible and interconnected, this division needs to be functional division not physical division. The following points summarize the reasons behind going into a layered model:

1. To simplify understanding the network model.
2. Layering the network based on functions makes it easier to implement, because the functions of each layer are distinct and consistent. Preparing the right software and hardware based on functions is much easier.
3. To simplify the troubleshooting of the network. With each layer's functions being distinct, the problem can be easily isolated and the error can be corrected without disturbing other network functions.
4. Layering the network makes it easier to develop. Development goes better and more focused when it goes in separate modules and protocols. When each layer has its own protocols, this layer's duties can be focused on and the protocols designed for this specific layer can perform their duties in a more efficient way.
5. The layered model guarantees better consistency in functions and protocols.

Chapter 2
OSI Model

Abstract This chapter starts with a brief history of the OSI model and how it all started in the mid 1970s. Afterwards, the OSI model is explained in details along with the functions and duties of each layer in the model. Studying the OSI model is a simple first step into the networking world. At the end of the chapter, the encapsulation and decapsulation processes are introduced such that the reader would understand the end-to-end data flow from one host to another.

Keywords OSI · ISO · Seven layers · Physical layer · Datalink · Network · Transport · Session · Presentation layer · Application layer

2.1 History of OSI Model

The OSI model was officially adapted as a standard by ISO in 1979. Some might say that it is an old standard. Well, it is old. What kept this model alive for so long is its capacity of expansion to meet the evolving needs.

Most of the work that created the base for the OSI model was done by a group at Honeywell Information Systems. The head of this group was Mike Canepa. This group started addressing the lack of standardization problem in the mid 1970s of the past century and they came up with a proposal named Distributed Systems Architecture, DSA. By that time, the British Standards Institute submitted a proposal to the ISO saying that there is a need for unified standard communication architecture for distributed processing systems. Responding to this proposal, the ISO formed a subcommittee on Open System Interconnection. The ISO also made American National Standards Institution (ANSI) in-charge of preparing proposals prior to the first official meeting of the subcommittee. Canepa's group participated in the ANSI meetings to discuss their seven-layer proposal. Later, ANSI chose to provide Canepa's proposal as the only one to be submitted to the ISO subcommittee.

In March 1978, the first meeting of the subcommittee was made and Canepa and his team presented their proposal there. The ISO group thought that this proposal covered most of the needs for Open System Interconnection. In the same

M. M. Alani, *Guide to OSI and TCP/IP Models*, SpringerBriefs in Computer Science, DOI: 10.1007/978-3-319-05152-9_2, © The Author(s) 2014

month that year a provisional version of the model was published. With some minor improvements, the next version of the model was published in June 1979 and was standardized.

In 1995 the OSI model was revised to cover the needs arising by the rapid development in the field of computer networks [1].

2.2 OSI Layers

The ISO OSI model consists of seven layers. Figure 2.1 shows these layers. Usually, the routers and other network devices act in the bottom three layers and the hosts act in the whole seven layers.

Each layer handles the data in a way that is different from other layers. The unit in which a certain layer handles data is called a Protocol Data Unit (PDU). Some layers add layer-specific information to the data. This information added by the layers' protocols can be in the form of a header, a trailer, or both. The header information is added at the start of the PDU, while the trailer information is added at the end of the PDU. This header or trailer contains information that is useful in controlling the communication between two entities.

The OSI model works in a peer-layer strategy. This strategy implies that the control information added to the PDU by one layer is meant to reach the peer layer in the receiving entity. For example, the header information added at the network layer in the sender host is used by the network layer in the receiving host and this

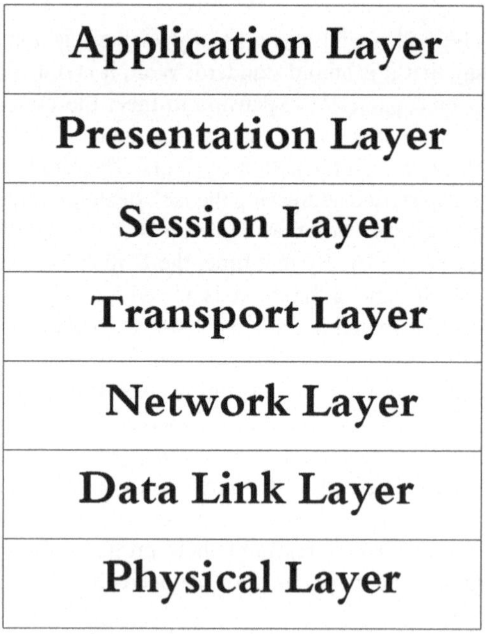

Fig. 2.1 The OSI-model
seven layers

| Application Layer |
| Presentation Layer |
| Session Layer |
| Transport Layer |
| Network Layer |
| Data Link Layer |
| Physical Layer |

information is insignificant to other layers. Hence, compatible protocols must be used at both ends of the communication to succeed in delivering the user data in the right manner.

Before going into a brief description of each layer, we need to add two new concepts. These concepts are the modes of data transfer; *connection-oriented* and *connectionless*. In connection-oriented communication, a connection needs to be established before the start of transmitting data from the sender to the receiver. This is analogous to a phone call. You cannot start talking to the other side of the phone before a connection between you and them is established and the other end actually picks up the phone. Connectionless communications refer to the type of communication in which there is no connection establishment before the transmission of data takes place. Control information is added to the data and the data is then sent to the destination and you cannot tell, in an easy way, whether the receiver has received the data correctly or not. This is analogous to sending a written message by mail. All you can do is write the address on the message and drop it at the post office.

The idea of the model system is not to tell you how the network actually operates, but to define the elements and functions that compose a network in a way that makes these elements and functions distinct and distributable on layers. This distinction provides the ability to protocols to operate in a smooth way and to be easy to troubleshoot.

In the following subsections, we will discuss each layer's functions and how each layer handles the data. In the next section, we will go through the complete cycles of data from source host to destination host [2].

2.2.1 Physical Layer

The physical layer basically handles data as raw bits. This means that the PDU for the physical layer is a *bit*. The primitive duty of the physical layer is to provide transparent transmission of bits from the data-link layer of the sender to the data-link layer of the receiver. This is accomplished by defining the mechanical, electrical, functional and procedural means to activate, maintain, and deactivate a physical link between two data-link entities.

Beside the data transmitted from one physical entity to another, control information needs to be transferred too. This control information may be added to the data and transformed in the same channel in which the data is transferred, and this is called *in-line signaling*. Or, the control information may be transferred through a separate control channel, which is called *off-line signaling* or *out-of-line signaling*. The choice of which way to transfer the control information is left to the protocol used.

Physical layer protocols vary depending on the type of the physical medium and the type of the signal carried on it. The signal can be an electrical voltage carried over a cable, a light signal carried through a fiber link, or even an electromagnetic signal carried in the air on in the outer space.

The main functions of the physical layer are:

a. Physical connection activation and deactivation.
   The physical connection activation and deactivation is done upon request from the data-link layer.
b. PDU Transmission.
   As we have mentioned before, the physical layer PDU is bit. So, transmission of bits from the source to the destination is a physical layer function.
c. Multiplexing and demultiplexing (if needed).
   There are many cases in which two or more connections need to share the same physical channel. In this case, multiplexing these connections into the channel is required at the sender side, and demultiplexing is required at the receiver side. This function is usually done through a specialized data-circuit, and is optional in the OSI standard.
d. Sequencing.
   The physical layer must make sure that the transmitted bits arrive in the same sequence in which they were sent from the data-link layer.
e. Physical layer management.
   Some layer management aspects are left to the protocol and used medium, such as error detection. These management functions depend on the protocol and physical medium. For example, the electrical signal transmitted through a metallic wire needs different management than the optical signal transmitted through a fiber cable.

2.2.2 Data-Link Layer

The PDU of the data-link layer is a *frame*, which means, the data-link layer handles data as frames. These frames may range from few hundred bytes to few thousand bytes. The data-link layer adds its control information in the form of a header and a trailer.

Data-link layer has many complex functions as compared to other layers. Data-link layer provides different type of functions for connection-oriented and connectionless communications. Actually, all the functions provided to the connectionless communication are provided to the connection oriented, but the opposite is not true. The following is a list of functions provided for both connection-oriented and connectionless communications:

a. Control of data-circuit interconnection
   This function gives network entities the capability of controlling the interconnections of data-circuits within the physical layer.
b. Identification and parameter exchange
   Each entity needs to identify itself to other entities and some parameters governing the communication need to be exchanged, too. An example of these parameters is data rate.

c. Error detection

Some physical channels might be susceptible to factors that prevent the data from being delivered in the right way. These factors can be Electro-Magnetic Interference (EMI), temperature, rain, etc., depending on the medium type. One of the data-link layer functions is to detect these errors.

d. Relaying

Some network configurations require relaying between individual local networks.

e. Data-link layer management

Similar to the physical layer management, the data-link layer leaves some management operations to the protocols used.

In addition to the functions listed above, the data-link layer provides the following functions only to the connection-oriented communications:

a. Data-link connection establishment and release

As the name indicates, this function is responsible for the establishment and release of the data-link connections between communicating entities.

b. Connection-mode data transmission

Connection-oriented communication requires certain mechanisms in order to assure the delivery of data. For example, in connection-oriented communication, for each transmitted frame, or group of frames, an acknowledgement frame is transmitted back from the receiver to the sender to acknowledge the reception of the frame or frames.

c. Data-link-connection splitting

This function is aimed to split the data-link connection into multiple physical connections, if possible.

d. Sequence control

This function assures that the data frames are received in the same order in which they were sent or at least assure that the frames can be re-arranged in the right order if they arrive out of order.

e. Framing (delimiting and synchronization)

This function provides the recognition of a sequence of bits, transmitted over a physical connection, as a data-link frame.

f. Flow control

In connection-oriented communication, the sender and receiver can dynamically control the rate in which the data is transferred. In connectionless communication, there is service boundary flow control but no peer flow control. This means that in connectionless communication, there is a limit imposed by the physical medium and physical layer protocol to the flow, but the rate can be controlled by the communicating entities.

g. Error recovery or error correction

This function tries to correct the detected error based on mechanisms used by the data-link protocol. In connectionless communication the data-link layer can only detect errors, but not correct them. This function tries to correct the error, and if it fails, it informs the network entities of that error to perform retransmission.

h. Reset

This function forces the data-link connection to reset.

2.2.3 Network Layer

The PDU for the network layer is a *packet*. The network layer handles very crucial duties regarding the routing of data from one network to another and controlling the subnet. Routing can be a complex operation in some times as many factors contribute in the choice of the best route from a source to a destination. The following is a list of the network layer functions:

a. Routing and relaying
   Routing is the operation of selecting the best path for data from source to destination and sending the data along that path.
b. Network connection and multiplexing
   This function provides network connections between transport-layer entities by employing the data-link connections available. Sometimes multiplexing is needed to optimize the use of these data-link connections by sending more than a single network connection through the same data-link connection.
c. Segmentation and blocking
   The network layer may segment and/or block the PDUs in order to facilitate the transfer. Segmentation, or sometimes referred to as Fragmentation, is basically making the PDUs smaller. This is an important function if the data is passed between networks that are using different data-link layer standards like Ethernet and Asynchronous Transfer Mode (ATM). These different data-link standards can have different maximum packet size. And thus causing the PDU of one data-link protocol incompatible with another data-link protocol.
d. Error detection and recovery
   In order to check that the quality of service provided over a network connection is maintained, error detection function is required. Network layer uses error notifications from the data-link layer and additional error detection mechanisms. The error recovery is also essential to try to correct the detected errors.
e. Sequencing and flow control
   Sequencing is used to maintain the sequential order of the packets sent to a destination upon the request of the transport layer. Flow control is used for prevent flooding the destination with data, and control the transmission rate.
f. Expedited data transfer

This function provides expedited data transfer from source to destination, if required.

g. Reset

Reset the network connection.

h. Service selection

This function assures the use of the same service at the source and destination as the packets pass through different subnetworks with different quality levels.

i. Network-address-to-datalink-address mapping

This mapping is important to facilitate the proper transfer of data from the source to the networking devices and from the networking devices to the destination, and back.

j. Network layer management

This function is to manage the network layer functions and services.

The network layer requires some facilities to be able to accomplish its functions. The facilities can guarantee the smooth operation of the layer. The most important of these facilities are network addressing (sometimes referred to as logical addressing), quality of service parameters, expedited PDU transfer, and error notification.

Network addressing gives a unique address to each host, thus, it is consistent. Hosts lying in the same subnetwork have to have a common portion of their addresses which is called a network address or a subnet address. To understand this easier, think of calling a phone number in another country. The format of the phone number is country-code—area-code—subscriber code. For example, the telephone number of a person in Virginia, USA should be in the format +1-703-XXXXXXXX. The first part, (1), identifies the country; USA. The second part, (703), identifies the area; Virginia. The third part identifies the subscriber within Virginia, USA. This hierarchy in telephone numbers is analogous to network addressing. A common part identifies the network, or subnet, address, and a unique part that identifies the particular host.

The quality-of-service (QoS) parameters define the quality limits of the network connection. Most known QoS parameters are delay, jitter, service availability, reliability, and network-connection establishment delay. These parameters are beyond the scope of this brief.

Error notification might lead to the release of the network connection, but not always. This depends on the specifications of the particular network service in which the error has been detects. Unrecoverable errors detected by the network layer are reported to the transport entities.

2.2.4 Transport Layer

Since there are two types of services that can be provided to the networking applications, connection-oriented and connectionless, the transport layer provides different kind of functions for these two types. The PDU for the transport layer is a segment.

The functions of the transport layer in a connection-oriented communication are listed in the following:

a. Establishment and release of transport connections
   This function is responsible for initiating the connection between the communicating entities and releasing the connection when the data transfer is over.
b. Sequence control
   Controlling the sequence of data transferred to guarantee that the data arrive in the same sequence in which it was sent.
c. End-to-end error detection and recovery
   This function provides detection of errors occurring in segments and trying to recover these errors to their original error-free form.
d. Segmentation
   At the transport layer, the data is transformed into segments at the sender and reconstructed at the recipient.
e. End-to-end flow control
   This function controls the rate in which segments are transferred from one entity to another.
f. Monitoring QoS parameters
   This function provides the transport layer the ability to monitor the QoS parameters of the communication.

For connectionless communications, the functions of the transport layer are:

a. End-to-end error detection
   In connectionless communications, the transport layer only detects the errors and notify the session entities, but does not try to recover them.
b. Monitoring of QoS parameters
   Connectionless communications can also be monitored in terms of QoS parameters.
c. PDU delimiting
   This function introduces the ability to delimit the PDUs to maintain the continuity of communication.

The transport layer gives a great support to the session layer in terms of providing the mechanisms to differentiate which data goes to what session.

The connection-oriented communication goes into three phases in the transport layer; establishment, data transfer, and release. During the establishment phase, the transport layer sets the parameters of end-to-end communication. For example, multiplexing of sessions into a network-connection, optimum segment size, and obtaining a network connection that matches the needs of the session entities. After establishing the connection, the data transfer starts and uses error detection and correction, sequencing, segmentation, and flow control mechanisms. When the data is transferred completely, the sender notifies the recipient of the request to release the connection, and the connection is released.

2.2.5 Session Layer

The session layer does not have a PDU of its own. It handles the data in the shape they come in, without division or concatenation. Its basic purpose is to provide the ability to the presentation entities to organize the communication for multiple communication sessions taking place at the same time.

The main functions of the session layer are:

a. Session initiation and teardown
   The session layer is responsible for starting the sessions between the communicating entities. And when the session is over, the session layer is responsible for the release of the communication. Data transfer takes place in-between.
b. Token management
   This function is related to the communication mode used in the specific session (simplex, half-, or full-duplex). The session layer controls which entity own the token and can transmit data at this time. The token is the license to transmit in an environment or service where only one entity can transmit at a time. This is not the case for all applications. Some applications operate in full-duplex mode, others operate in half-duplex mode.
c. Session-connection to transport-connection mapping (in connection-oriented transfer only)
   The function provides the session layer the ability to map between the transport layer connections and the sessions currently taking place. This way the session layer can tell which data goes to what session.

2.2.6 Presentation Layer

The presentation layer, like the session layer, does not have a PDU of its own. The presentation layer, as its name indicates, is responsible for the way data is presented to the application. At the start of the communication, the presentation layer negotiates the form of data to be transferred with the other entity, the *transfer syntax*. After this negotiation, the presentation layer can provide additional services such as compression, encryption, and translation. The choice of which service(s) to be used is up to the application itself.

2.2.7 Application Layer

The application layer is responsible for defining the services presented at the user-end. Application layer protocols vary according to the specific type of data the user wants to transfer. Application layer also defines the acceptable QoS parameters for each service. For example, the voice data transmission requires different QoS parameters than transferring an email. The choice of which security aspects to use, such as authentication and access control, is up to the application layer. Also the synchronization of communicating applications in connection-oriented services is the application layer's responsibility.

Generally, the main functions of the application layer are:

a. identification of services provided to the user
b. defining QoS parameters required by the application
c. defining security mechanisms to be employed such as access control and authentication
d. synchronization of communicating applications (only in connection-oriented services).

2.3 End-to-End Data Flow

The flow of data from the application layer to the physical layer is called *encapsulation*. This is because header and trailer information is added to the data in various layers at the end and start of data, which makes it look like a capsule. The flow of data in the opposite direction, from the physical layer to the application layer, is called *decapsulation*, as it involves the removal of the headers and trailers such that the data is back to its original form to the receiver's user-end. Figure 2.2 shows the details of the encapsulation process.

Figure 2.3 shows the process of transferring data from on host to another. Since the OSI principle is to transfer data between different networks, not transferring data within the same network, we added a router in the middle between the two communicating hosts. In real life, there can be more than one router, depending on the specific networks we are dealing with. All routers act in a fairly similar way with the data received. So, we can replace the router in the middle with a series of routers.

2.3.1 Host A (Source)

1. The transmission starts from the application layer at Host A. The application decides that it needs to communicate with Host B and passes the data down to the presentation layer.
2. The presentation layer does the required transformations that need to be done on data, like compression, encryption, or translation. Then the data is passed down to the session layer.

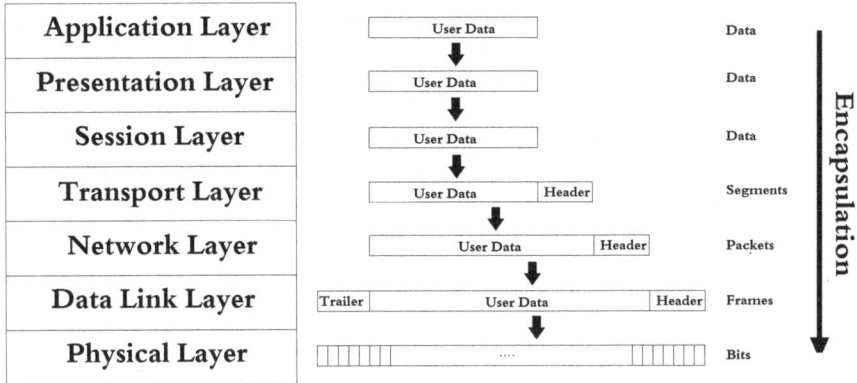

Fig. 2.2 Data encapsulation process in OSI model

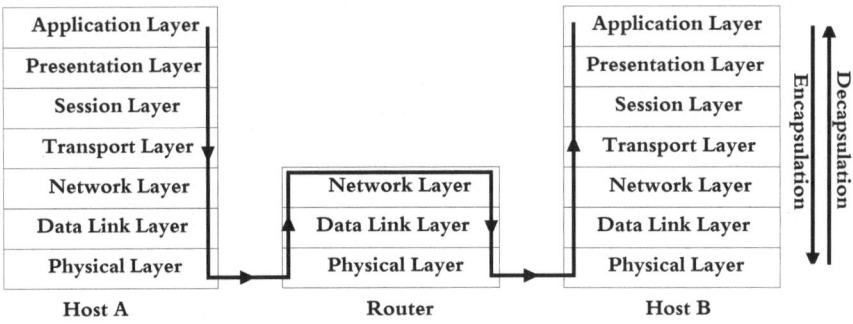

Fig. 2.3 End-to-end data flow in OSI model

3. The session layer starts initiating the communication session and passes the data to the transport layer.

4. At the transport layer, the data is segmented and a header is added to each segment of the data. This header contains transport control information such as the sequence number and acknowledgement number. The segment along with its header is passed down to the network layer.

5. The network layer deals with the whole segment, including its header, as data. The header added by the transport layer is meant to be read by the transport layer at the receiving end. So, the network layer does not read the segment header, but instead, it handles the segment and its header as a single data element. The data is then put into packets and headers are added to these packets. The network layer header contains information meant to reach the network layer on the other end. This information includes a source and destination network-layer address along with few other parameters. These packets are sent down to the data-link layer.

6. The data-link layer handles the packet and its header as a single data element, such that the network layer header is considered part of the data. The data-link layer puts the data into frames and adds a header and a trailer to each frame. The data-link header contains control information sent to the data-link layer on the other end, while the trailer usually contains error control information.
7. The frames are then sent down to the physical layer where they are dealt with as raw bits. These bits are transferred through the physical channel to the router.

2.3.2 The Router

The router does a partial decapsulation because it does not need to read the data all the way up to its application-layer shape. The router needs only to read up to the network layer header to route the data to the wanted destination.

1. The process starts at the physical layer when the data is received as raw bits. These bits are then gathered into frames and sent over to the data-link layer.
2. The data link layer reads the header and the trailer to know what to do with the data. The data-link layer then rips off the header and the trailer and sends the rest of the data as a packet to the network layer.
3. At the network layer, the network header is read to determine the destination network address. After knowing the destination network address, the router chooses the best route to send the data to the destination host.
4. Starting from the network layer, the encapsulation starts again at the router. Data goes down the data-link and physical layers and all the way through the physical link to the destination host.

If there is another router in the way, the same process that took place in the first router is repeated until reaching the destination host.

2.3.3 Host B (Destination)

1. At the destination host, the raw bits are elevated as frames to the data-link layer.
2. The header and trailer of each frame are read by the data-link layer and then removed. The rest of the data is elevated to the network layer as packets.
3. The header of the packet is read to determine if this is the correct destination for the packet and other network control information is also taken from the network layer header. The header is then removed and the rest of the data is elevated to the transport layer as segments.
4. The header of each segment is read to determine the sequence number and arrange the segments in their correct order. Then, the header is also removed

and the rest of the data is elevated to the session layer. The transport header also contains information of which session this data is going to. This information is passed to the session layer with the data.

5. The session layer determines if this is the end of this session or not. If it is the last segment in the session, the session layer will wait for the request to end this session. If this is not the last segment in the session, the session layer waits for more data.

6. The data is then passed to the presentation layer to retransform the data into the shape that they were sent in by the sending-end application or to another form determined by the application. This might involve decompression, decryption, or translation.

7. The data is then transferred to the application and received by the user.

One important thing to remember is that each layer, using its header, trailer, or connection-setup parameters, communicates with the peer layer at the receiving end.

References

1. Stallings, W.: The Origins of OSI [Online] (1998). http://www.williamstallings.com/Extras/OSI.html
2. ISO: Information Technology—Open Systems Interconnection—Basic Reference Model: The Basic Model. Geneva, Standard ISO/IEC 7498-1(E) (1994)

Chapter 3
TCP/IP Model

Abstract This chapter gives a brief introduction to the TCP/IP model and the protocols used in each layer. The chapter starts by a historical background about the TCP/IP model and how it all started. The TCP/IP model layers are briefly explained in contrast with the OSI model and its layers. The network access layer is introduced and the flexibility of the TCP/IP model in this layer is also explained. Then, the internetwork layer is explained with brief description of the most important protocols operating in the network layer; IP, ICMP, ARP and RARP, and InARP. The next section provides brief description about the transport layer and the two protocols operating in this layer; TCP and UDP. The last section of this chapter describes the application layer of the TCP/IP model. A brief description of the most commonly used application protocols; HTTP, DNS, FTP and TFTP, SMTP, POP3, and Telnet, is also given.

Keywords TCP · IP · HTTP · FTP · POP · POP3 · TFTP · Telnet · Application layer · Internetwork · Standard · TCP/IP · ARP · RARP · Inverse ARP

3.1 History of TCP/IP

The TCP/IP model was a successor of an older project by the US Department of Defense (DoD) called Advanced Research Projects Agency NETwork (ARPANET). ARPA was later called Defense Advanced Research Projects Agency (DARPA). *ARPANET* was a communications network that is designed to survive the loss of communication subnet such that the communication continues as long as the source and destination of the conversation exist. This project first saw the daylight in 1969.

When new communication technologies evolved, such as the wireless technologies, the need for a newer model evolved. ARPANET was then developed to the TCP/IP model. Although the first definition of the TCP/IP model was introduced in [1] in 1974, ARPANET did not completely adopt it until 1983. Starting from that year, the ARPANET was gradually called the *Internet*.

M. M. Alani, *Guide to OSI and TCP/IP Models*, SpringerBriefs in Computer Science, DOI: 10.1007/978-3-319-05152-9_3, © The Author(s) 2014

The TCP/IP model took its name from the two essential protocols that create together the backbone of the model. Unlike the OSI model which was created as a model so that protocols are developed based on it, the TCP/IP model was based on protocols that already existed, or earlier versions of them existed. This gives more flexibility to the OSI model over the TCP/IP. Still, the TCP/IP was world widely adopted because the ARPANET was already there and the ARPANET adopted it. Despite that, studying the OSI model gives a very educational insight about the way networks operate, or should we say, inter-operate. The TCP/IP was widely spread for a various number of reasons, beside the fact that it was already there. The hardware and software independency, the large and flexible addressing scheme, and free availability and public documentation of the standards were also important reasons that lead to the wide adoption of the TCP/IP standard [2].

3.2 TCP/IP Layers

Unlike the OSI model, the TCP/IP model consists of four layers; Network Access layer (or sometimes called Host-to-Network layer), Internetwork layer (sometimes known as Internet layer), Transport layer, and Application layer. These four layers are shown in Fig. 3.1.

The first thing noticed when looking at Fig. 3.1 is that there are no session or presentation layers, and the data-link and physical layers are reduced to one layer. The layers of the TCP/IP model will be discussed in the coming sections.

The TCP/IP model was created based on a certain set of protocols. Unlike the OSI model which was created as a layered model first with clear defined functions. Thus, a better understanding of the TCP/IP will be in a better understanding of how the protocols in its protocol set works. Figure 3.2 shows the TCP/IP model layers in contrast to the OSI model layers.

3.3 Network Access Layer

This is not an actual layer, in the broad sense of a layer. The TCP/IP standard does not discuss the details of this layer. The duty of this layer is to make sure that the IP packets coming from the Internetwork layer are delivered into a physical link and on the other side, and the opposite is done.

This is one of the points of strength of the TCP/IP model. The model does not care what type of local or wide area networking technology is used or what type of medium, as long as this network is able to deliver IP packet. This means that LAN and WAN technologies such as Ethernet, Fiber Distributed Data Interface (FDDI), Frame Relay, and Wireless technologies of many kinds can be used below the Internetwork layer.

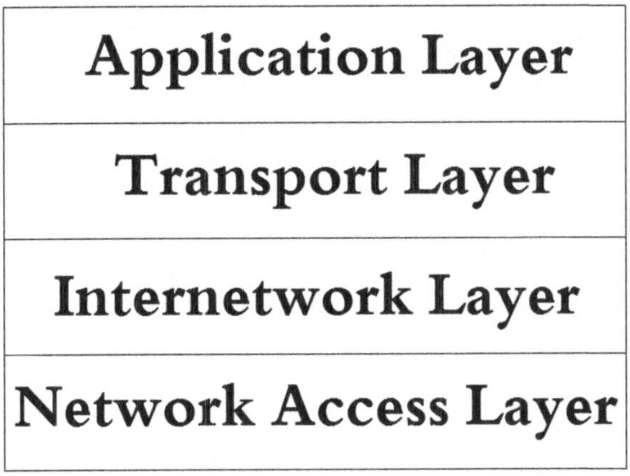

Application Layer
Transport Layer
Internetwork Layer
Network Access Layer

Fig. 3.1 TCP/IP model layers

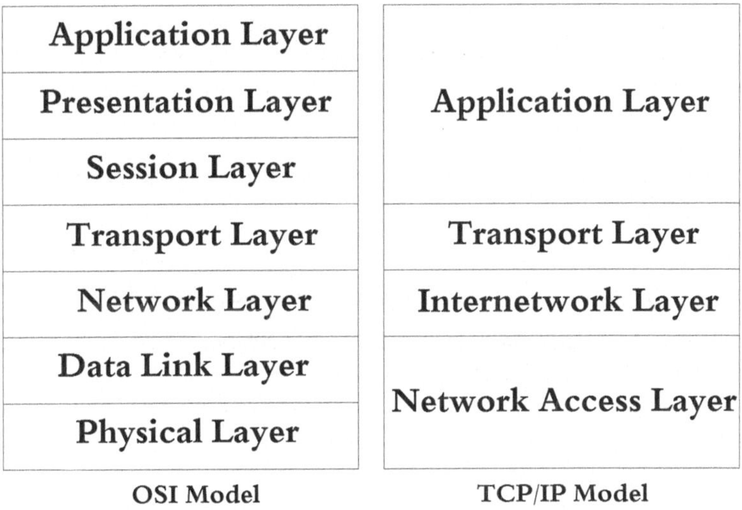

OSI Model	TCP/IP Model
Application Layer	Application Layer
Presentation Layer	
Session Layer	
Transport Layer	Transport Layer
Network Layer	Internetwork Layer
Data Link Layer	Network Access Layer
Physical Layer	

Fig. 3.2 OSI and TCP/IP reference models: **a** OSI model, **b** TCP/IP model

Basically, protocols operating in this layer should define the procedures used to interface with the network hardware and access the transmission medium. Mapping the IP addresses used in the Internetwork layer to hardware addresses (such as the MAC address) is yet another duty of this layer. Based on the type of the hardware and network interface, the network access layer defines the physical media connection.

Most of the work done in this layer is done by the software and drivers assigned to handle the networking hardware. In most cases, the configurations that need to be done are simple, such as installing or activating the TCP/IP software stack. In most cases the required software is preloaded in most computers such that the user can plug into some networks directly. In other cases, further configuration steps are required to get connected, such as configuring IP addresses, subnet masks, and gateway addresses.

This layer differs from other TCP/IP layers in that it makes use of the existing LAN and WAN standards rather than defining its own new standards.

3.4 Internetwork Layer

Sometimes this layer is called the *Internet Layer* or the *TCP/IP Network Layer*. The main purpose of this layer is to select the best path for the data to travel through from the source to the destination, i.e., *routing*. The leading protocol operating in this layer is IP. There is a group of supporting protocols that support IP in doing its job, such as the Internet Control Message Protocol (ICMP), Address Resolution Protocol (ARP), and Reverse Address Resolution Protocol (RARP).

Routing is the process of selecting the best path for a packet to reach its destination.

The main operations done by the IP protocol as defined by RFC791 are:

1. Define a packet (datagram) and an addressing scheme.
2. Transport data between network access layer and transport layer.
3. Fragment and reassemble packets at the source, routing hops, and destination. Reassembly happens at the destination only.
4. Choose the best route for data from source to destination.

The IP protocol is a connectionless protocol. This means that IP protocol does not establish a connection before the transfer of information and does not employ an acknowledgment mechanism to assure the delivery of its packets. IP protocol leaves these operations to protocols of higher layers, such as the TCP. IP protocol also relies on other layers to perform error detection and correction. Thus, IP protocol is sometimes referred to as unreliable. This does not mean that the IP protocol can not be relied on in delivering packets safe and sound to the other end; it just means that the error handling and connection setup is left for other protocols to handle. Based on that fact, IP protocol is often defined as an unreliable best-effort delivery protocol.

3.4.1 The Internet Protocol

IP protocols is a packet switching protocol defined by the IETF RFC971 and was amended later by the RFCs 950, 919, and 922. This protocol was designed to work in interconnected systems of packet-switched computer communication networks. The main duty of this protocol is to deliver packets from one host in a network to another host laying in the same or a different network. This is achieved by adding a header that contains addressing and control information. This header contains a source and a destination address that are defined as *IP addresses*. An IP address is a 32-bit address in a format called *dotted-decimal* (for example 192.168.0.1). The dotted decimal format divides the 32 bits into four 8-bit groups, and each of these four groups are turned into decimal format with a dot separating one group from the other.

IP protocol provides segmentation and reassembly of lengthy packets into smaller packets. This becomes very useful when the packets go through networks that have different rules of maximum packet length on the way to the destination network. IP protocol treats each packet as an independent entity unrelated to any other packet. There are no connections or logical circuits (virtual or otherwise). Figure 3.3 shows the anatomy of the IP packet [3].

The *VER* field is used to identify the version of IP protocol used, whether it is IPv4 or IPv6.

The fields *HLEN* and *Total Length* identify the length of the IP packet header and the total length of the IP packet, consecutively.

Fragment Offset and part of the *Flags* fields are used to facilitate the fragmentation process that happens when the packet is too large to pass in a network on its way to the destination.

Source IP Address and *Destination IP Address* fields are used to identify the unique logical addresses of the source and destination of packets. These unique addresses are used by the routing hops to identify the best path to deliver the packet to its destination.

Using its header information, IP protocol uses four key mechanisms in providing its services to other layers; Type of Service (ToS), Time to Live (TTL), Options, and Header Checksum as defined in IETF RFC 791.

The *Type of Service* field is used to indicate the quality of the service desired. The type of service is a generalized set of parameters which characterize the service quality choices. This type of service indication is to be used by gateways in the route to the destination to select the actual transmission parameters for a particular network, the network to be used for the next hop, or the next gateway when routing a packet. In general, the standard identifies five different types of services; minimize delay, maximize throughput, maximize reliability, minimize monetary-cost and normal service. Each type tells the networking device how this packet should be handled. Table 3.1 shows the default values of ToS field for different application layer protocols in different operation modes.

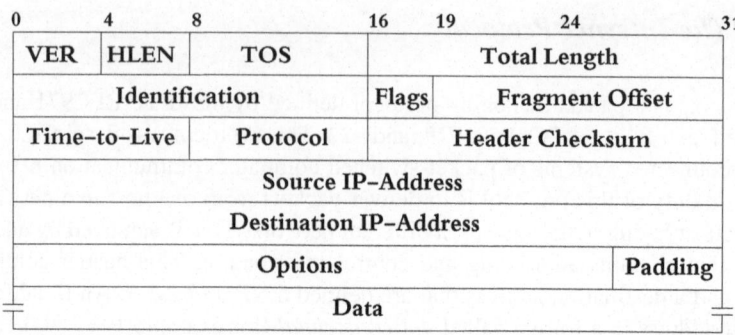

0	4	8		16	19	24	31
VER	HLEN	TOS		Total Length			
Identification				Flags	Fragment Offset		
Time–to–Live		Protocol		Header Checksum			
Source IP–Address							
Destination IP–Address							
Options						Padding	
Data							

Fig. 3.3 Anatomy of an IP packet

Table 3.1 Default ToS values for different application layer protocols

Protocol	Minimize delay	Maximize throughput	Maximize reliability	Minimize monetary cost	Normal service
FTP control	√				
FTP data		√			
TFTP	√				
Telnet	√				
SMTP command	√				
SMTP data		√			
DNS UDP query	√				
DNS TCP query					√
DNS Zone transfer		√			
EGP routing protocols					√
IGP routing protocols			√		

The *Time to Live* (TTL) is similar to setting an expiry time for the packet. This way it is guaranteed that packets that are lost do not keep wandering around the network looking for their destination. In simple words, TTL is the number of routing nodes (or hops) the packet is allowed to pass through before it gets discarded. It is set by the sender of the packet as the maximum number of hops the packet can pass through from the source to the destination. This TTL is a down-counter that is reduced by one at each hop along the route where it is processed. If the TTL reaches zero before the packet reaches its destination, the packet is discarded. The TTL can be thought of as a self destruction timer.

In a world where about 2.4 billion people use the Internet, according to (Internet-World-Stats 2012), if each user send one lost packet every minute, we would have 24 billion lost packet within ten minutes only. If TTL did not exist, the Internet would've been down within a few minutes.

The *Options* field contains control functions needed in few situations but unnecessary for the most common communications. The options include provisions for timestamps, security, and other special routing needs. This field can be used to include source routing, and route recording.

The *Header Checksum* provides verification that the header information used in processing the packet has been transmitted correctly. However, the packet's data may contain errors. If the header checksum fails, the packet is discarded at once by the entity which detects the error whether it is the destination or a gateway in the route.

As stated earlier, the IP protocol does not provide a reliable communication facility. There are no acknowledgments neither end-to-end nor hop-by-hop. There are no error detection and correction mechanisms for the data, only a header checksum that checks for bit errors in the IP header. There are no retransmission or flow control mechanisms. IP relies on ICMP in reporting the detected errors to the source.

3.4.2 Internet Control Message Protocol

IP is used for host-to-host packet service in a system of interconnected networks. The network connecting devices, i.e. gateways, communicate with each other for control purposes via a special gateway to gateway protocol. From time to time, a gateway or destination host needs to communicate with a source host (for example, to report an error in packet transmission). For such purposes this protocol, ICMP is used. ICMP uses the basic support of IP as if it were a higher layer protocol; however, ICMP is considered an integral part of IP [4].

ICMP uses the regular IP packet to transfer its information. The typical ICMP message structure is shown in Fig. 3.4.

The *Type* field is used to define the type of the ICMP message, for example, error message. There are almost forty standard message types that were designed to accomplish multiple functions starting from echo request and reply (used in PING command), to destination unreachable (to report lost or undelivered packets), and many more.

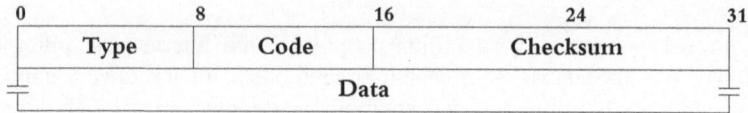

Fig. 3.4 Typical ICMP message structure

The *Code* field is used to define the subtype of the message within the specified ICMP message type. For example, message type 3 (Destination Unreachable) has 16 different subtypes like destination network unreachable, destination host unreachable, destination protocol unreachable, ...etc.

The *Checksum* field is a 16-bit checksum that is calculated in a manner similar to the IP header checksum. While the IP header checksum is used to detect errors in the IP header only, the ICMP checksum provides bit error detection coverage for the entire ICMP message.

Mainly, the contents of the *Data* depend on the type and code of the ICMP message. For some messages, the Data field contains the IP header and the first 64 bits of the packet that triggered the error that caused the ICMP message to be sent. In certain cases, such as the echo request or reply, data field contains three timestamps; originate, receive, and transmit timestamps.

The main duties of ICMP are the following:

1. Report network errors such as a host or a network being unreachable.
2. Report network congestion. If the buffers of a router or a gateway are full, and more and more packets are being received, the router or gateway sends a congestion notification to the sender or to the previous router or gateway.
3. Support troubleshooting using the echo request and reply messages. This is the mechanism used in the ping and trace commands.
4. Report timeouts. When the TTL of an IP packet reaches zero without reaching its destination, the packet is dropped. When the packet gets dropped, the router or gateway that dropped it reports back to the packet source that this packet was dropped using an ICMP message.

3.4.3 Address Resolution Protocol and Reverse Address Resolution Protocol

The main purpose of the ARP protocol is to map between logical protocol addresses (for example IP address) and hardware local addresses (for example Ethernet MAC address). Although this protocol was originally designed for the 10 Mb Ethernet, it has been generalized to allow it to be used for other types of networks as defined in IETF RFC826 [5]. Figure 3.5 shows a typical ARP packet.

The *Hardware Type* field is used to indicate the hardware standard used. Ethernet uses the number 01. The *Protocol Type* field indicates the type of protocol used in the Internet Layer. The IP protocol uses the number 80. The *Hardware Length* field specifies the number of bytes used for the hardware address. For the Ethernet protocol, the length of the hardware address is 6 bytes (or 48 bits). The *Protocol Length* field is used to indicate the length of the protocol address in bytes. The IP protocol has a 4-byte (or 32-bit) address. The *Operation* field indicates whether this packet is a request or a reply. The request uses the value of 1, while the reply uses the value of 2. The *Source Hardware Address* field, as its name indicates, is used to contain the hardware address of the source. This field's length is usually 6 bytes for the Ethernet MAC address. The *Source Protocol Address* field contains the logical protocol address, usually the IP address, of the source. If the protocol is IP, this field's length is 4 bytes. The *Destination Hardware Address* field contains the hardware address of the destination. For the Ethernet protocol, the field's length is 6 bytes and it should contain the destination MAC address. The last field in the packet is the *Destination Protocol Address*. This field contains the destination logical protocol address. When the IP protocol is used, this field's length is 4 bytes and it should contain the destination IP address.

The typical scenario of the ARP usage goes as follows:

1. The source node has an IP address of a destination and wishes to know the hardware address so it can send the Ethernet frames to the destination.
2. The source sends an ARP request. Since the source does not know the destination hardware address, the destination hardware address is set to broadcast address. This way, all nodes in the local network receive a copy of the ARP request.
3. The destination node that has the IP address contained in the destination protocol address field responds with its hardware address.
4. The reply ARP packet contains the hardware and IP addresses of the node that sent the ARP request are used here as the destination addresses, while the hardware and IP addresses of the responding node are used in the reply as the source addresses.

The actual ARP protocol duty is the mediation between the internetwork layer and the network access layer. Thus, sometimes the ARP protocol is considered a supporting protocol in the network access layer, and some other times it is considered a supporting protocol in the internetwork layer.

The RARP protocol works in a similar way to that of ARP. The main difference is that RARP is used to supply the logical protocol addresses to the nodes that have only hardware addresses.

The anatomy of the RARP packets is very similar to that of the ARP packets. The main difference happens in the operation field. The RARP protocol uses the values 3 and 4 for request and reply respectively as indicated in IETF RCF 903 [6].

The main use of the RARP protocol is to provide IP addresses to the hosts who do not know their own IP address. The host sends a RARP request to the gateway, and the gateway looks into its ARP cache for the IP address related to this host's hardware address, and when a match is found the IP address is sent to the host.

0   8   16   24   31

Hardware Type		Protocol Type	
Hardware Len	Protocol Len	Operations	
Source Hardware Address (0–3)			
Source Hardware Address (4–5)		Source Protocol Address (0–1)	
Source Protocol Address (2–3)		Destination Hardware Adrs (0–1)	
Destination Hardware Address (2–5)			
Destination Protocol Address (0–3)			

Fig. 3.5 Typical ARP packet

RARP is mostly used to assign IP addresses to keyboard-and-monitor-only terminals that need to be connected to the network but do not have the ability to do it by itself.

3.4.4 Inverse ARP

Inverse ARP (InARP) is another supporting protocol that is not as widely used as the ARP and RARP. InARP is used with Frame-Relay networks to provide the job of ARP in FrameRelay interfaces. Basic InARP operates in a way similar to ARP with the exception that InARP does not broadcast requests [7].

Frame-Relay is a WAN technology that uses a virtual circuit identifier, Data Link Connection Identifier (DLCI), instead of using source and destination address. This DLCI number identifies the virtual circuit that connects the source to the destination. The main duty of the InARP is to provide mapping between logical protocol addresses and virtual circuits for the Frame-Relay network.

3.5 Transport Layer

The transport layer in the TCP/IP model has similar purposes to that of the OSI model. It is designed to give the source and destination the ability to have end-to-end conversation.

In the TCP/IP model, there are two defined protocols that can operate in this layer; TCP and UDP. These two protocols provide connection-oriented and connectionless communications. Only TCP protocol provides a way of sequencing, such that even if the data segments arrive in a different sequence of which they were sent in, they can be rearranged.

Since the IP protocol is unreliable, the use of TCP provides the reliability needed to assure that data arrives safe and sound. With the use of UDP the situation is different. UDP does not provide reliability mechanisms. Thus, many application layer protocols use TCP as their transport protocol.

Sometimes the overhead caused by the reliability provided by the TCP protocol compromises the quality of some time-critical communications like Voice over IP (VoIP) and video conferencing traffic. This leads to a conclusion that there is no such thing as a "better transport protocol". Each of the two protocols is used as the transport protocol for a wide scope of applications and conditions such that it is irreplaceable.

3.5.1 Transmission Control Protocol

TCP is one of the important building blocks in the TCP/IP protocol suite. As defined in RFC793, "TCP is a connection-oriented, end-to-end reliable protocol designed to fit into a layered hierarchy of protocols which support multi-network applications" [8].

In simple words, the TCP protocol establishes a connection between the source and the destination. During this connection, TCP protocol breaks data into segments at the source and reassembles them at the destination. Any segment that is not received, or received in error, is resent. Figure 3.6 shows a typical TCP segment.

The *Source Port* field contains a 16-bit port number. This port number is used to identify the session responsible of sending data at the source host. The *Destination Port* field contains a 16-bit port number that is used to identify the communicating session and distinguish the application that is expected to respond to this communication at the destination end. There are three ranges of port numbers; well-known ports (0–1,023), registered ports (1,024–49,151) and private ports (49,152–65,535). Ports are used by TCP as an interface to the application layer. For example, the FTP server is assigned, by default, to a well-known port of number 21. Each unique source and destination IP addresses along with unique source and destination port numbers define a single TCP connection that is unique, globally. This identifier (IP address:Port number) is sometimes referred to as socket. The *socket* is an interface created by the application and used to send and receive data through the network.

At the client side, a *client socket* is created to initiate the TCP session with the server. The source port number is assigned dynamically based on the available ports.

A socket is an application-generated,OS-controlled interface that a process uses to send and receive data through a network.

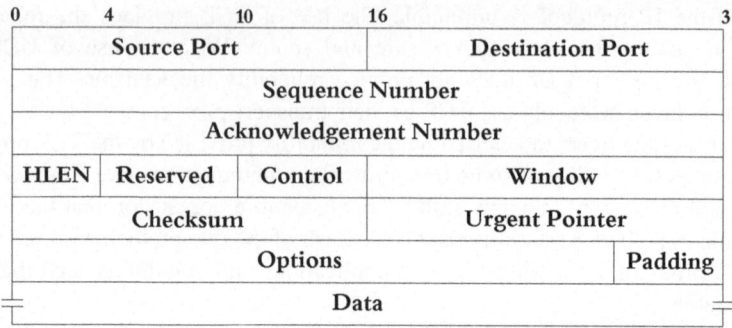

Fig. 3.6 Typical TCP segment

Each server has a *server socket* that is always open and waiting for requests to come from clients. After receiving the request from a client, a new socket is created to serve the client's request and to keep the server socket free to receive more requests.

The *Sequence Number* field contains a 32-bit number that defines the sequence number of the first byte carried in this segment in the whole byte stream for the TCP connection.

TCP sees user data as a stream of bytes. Each byte in this stream has a number. The idea behind this numbering is to give the receiving host the ability to re-arrange segments that arrive out of order, and to assure the reliability along with the acknowledgement number.

Each segment, from a large group of segments composing the user data, is assigned a sequence number. This number represents the sequence number of the first byte of user data carried in this particular segment. Figure 3.7 explains how sequence number is assigned.

The *Acknowledgement Number* field contains a 32-bit number that defines the sequence number of the next data byte expected to be received by the sender of this segment. Figure 3.8 provides a simple example of the concept of sequence number and acknowledgement number.

The example shown in Fig. 3.8 shows that each side of the communication has its own sequence numbers for the data it wants to transfer. Thus, A sends the acknowledgement number that is the sequence number of the last byte received from B plus one. While B sends acknowledgement number that is the sequence number of the last byte received from A plus one. Using this mechanism, the receiver will be well informed if a segment was lost on its way.

A common misunderstanding happens when someone thinks that the sequence number is the number of the segment. Actually, it is the number of the first byte in the segment, not the number of the segment itself.

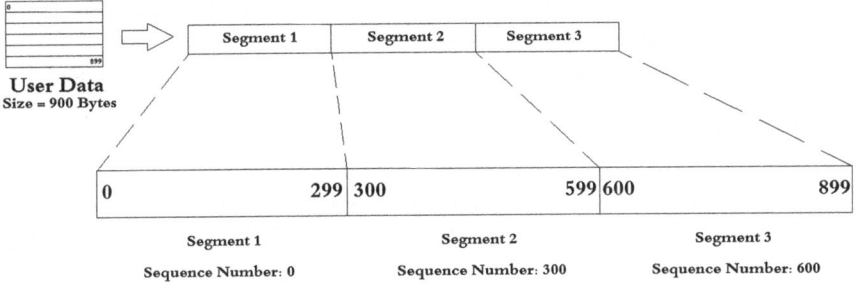

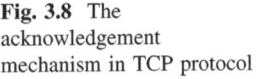

Fig. 3.7 Sequence number selection

Fig. 3.8 The acknowledgement mechanism in TCP protocol

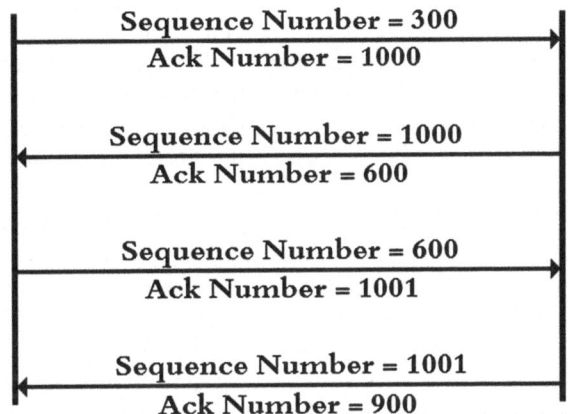

The *HLEN* or *Header LENgth* field in a TCP segment is a 4-bit number that indicates the length of the TCP header. The unit of this length is 32-bit words. Meaning that if HLEN is 5, for example, the header length is 5×32 bits in total. The 6 bits marked as the *Reserved* field are not currently used, but might be used in the future. A valid question would be why does the destination node need to know the length of the header? Isn't it a standard length? Well, the answer is that the length of the *Options* field may vary from one segment to the other. Thus, knowing the header length with enable the node to recognize where the header ends and data starts.

The 6 bits in the *Control* field are used as individual bits for the following purposes:

1. Urgent Pointer (*URG*): When this bit is set, the *Urgent Pointer* field should be interpreted by the receiving side.
2. Acknowledgement (*ACK*): When this bit is set, the *Acknowledgement* field in the TCP segment is valid.

3. Push Function (*PSH*): When this bit is set, the receiver is obligated to deliver this segment to the receiving application as soon as possible.
4. Reset the Connection (*RST*): When this bit is set, the receiver is informed that the sender is aborting the connection and all queued data and allocated buffers for the connection can be freely relinquished.
5. Synchronize (*SYN*): When this bit is set, it informs the receiver that the sender is attempting to synchronize sequence numbers. This function is usually required at the early stages of a connection establishment between the sender and the receiver.
6. Finished (FIN): This bit informs the receiver that the sender has reached the end of its byte stream for the current TCP connection so that the receiver shall not expect any more data.

The *Window* field is used to indicate the number of bytes the sender of this segment is willing to accept. Usually, the number is chosen based on the currently free buffer size, expected free buffer size, number of recently lost segments, and number of segments received recently with errors. This field is a part of the flow control mechanism used by TCP. This field defines how many bytes the sender should send before it pauses and waits for an acknowledgment from the destination node.

The 16-bit *Checksum* field is filled with checksum information for the TCP header and data. The receiving host generates a checksum of its own for that segment and compares it with the one that is sent in the segment. When the two checksums match, the receiver understands that the TCP segment has arrived with no errors. If the received and calculated checksums do not match, the segment is discarded and retransmission is required.

The *Urgent Pointer* field points to the end of the urgent data sent to the receiver. Sometimes the TCP sender needs to inform the receiver of the existence of urgent data that needs to be handed over to the application immediately. The content of the Urgent Pointer field is an offset number that is when added by the receiver to the sequence number of the same segment will point to the end of the urgent data. As mentioned before, this field is effective only when the *URG* bit in the *Control* field is set.

The *Options* field varies in length depending on the specific situation. Sometimes it is not used at all, and some other times two or more bytes are used. The main idea of the Options field is to add functionality to the TCP protocol such that it might be expanded for future uses or tailored for specific or new applications. The *Padding* is just a group of zeros added to the *Options* field so that the number of bits in the header is always a multiple of 32.

TCP employs a mechanism called *Positive Acknowledgments*. This means that only segments that are received in good condition with no errors will be acknowledged. There are no acknowledgements nor any notifications sent when a segment is lost or a segment is received with errors. Instead, the source node will setup a countdown timer for each segment sent. If a segment's timer expires before receiving an acknowledgment of this segment is received, retransmission is invoked.

Fig. 3.9 UDP segment format

0		16	31
Source Port		Destination Port	
Length		Checksum	
Data			

The method of calculating the initial value of this timer is beyond the scope of this brief.

3.5.2 User Datagram Protocol

The UDP protocol was designed to provide the applications with the ability to send data with minimal protocol overhead. The UDP is defined in [9] as "transaction oriented", and delivery and duplication protection are not guaranteed. Thus, UDP is considered a connectionless unreliable transport protocol. Figure 3.9 shows the UDP segment format.

The *Source Port* and *Destination Port* fields work in the same way they do in the TCP protocol; to identify the processes that are handling the communication and distinguish them from other sessions on the same host.

The *Length* indicates the total number of bytes in this UDP segment including the UDP header and the data. This means that the minimum value for the *Length* field is eight, because this is the length of the UDP header alone without data.

The *Checksum* field contains the checksum data calculated for the UDP header and the whole UDP segment along with parts of the IP header inside the data of the UDP segment. Filling this field is optional in UDP. Thus, not all nodes fill it up.

The lack of reliability and sequencing in UDP brings only one question in mind; does any application really use UDP? The answer is yes.

UDP protocol does not have connection establishment or teardown mechanisms. This can be considered a disadvantage for some application, and an advantage for other applications. Some applications need to send data immediately without waiting for the connection establishment procedure. This handshake procedure used in TCP, called Three-way Handshake, can take a noticeable amount of time that some application can not afford to lose.

Looking at Figs. 3.6 and 3.9, it is clear that the header of the TCP protocol over-sizes the UDP header with about 20 bytes in every single segment. This is not a small amount when transferring a large number of segments.

Sequence and acknowledgements monitoring, flow control, and other connection maintenance mechanisms used by the TCP protocol put extra load on the hosts on both ends of the connection as well as on the network itself. Thus, UDP protocol, with its less processing requirements makes it a light and suitable choice for real-time applications like VoIP and video conferencing. Real-time applications require timely transfer of data even if some segments are lost along the way.

One last thing that some applications find useful in the UDP protocol is the unregulated data rate. TCP protocol uses flow control mechanism such that it controls the rate in which the data is transferred from the sender. UDP protocol lacks this ability. Some applications need to take control of the rate in which they transfer data. Despite all that, reliability is an important issue that many applications require.

UDP is commonly used in multimedia applications such as video and audio conferencing, IP phones, and streaming of stored audio and video. These applications use UDP with a belief that they can tolerate a reasonable fraction of packet loss. The reaction of real-time video or audio streaming to the TCP congestions is very poor. That is why UDP is used in these applications.

3.6 Application Layer

The application layer is thought of as the crown of the TCP/IP model. As seen in Fig. 3.2, the TCP/IP model does not have session nor presentation layers. The designers thought that they were not that necessary. Thus, the application layer in the TCP/IP model handles data representation, encoding, and dialog control.

The main duty of this layer is to take data from the applications and deliver it to the transport layer and collect data from the transport layer and deliver it to the correct applications.

The TCP/IP model contains a huge group of high-level protocols that cover a wide range of applications. The most common is the Hyper Text Transfer Protocol (HTTP). Other widely used application layer protocols are File Transfer Protocol (FTP), Simple Mail Transfer Protocol (SMTP), Post Office Protocol 3 (POP3), Telnet, and Domain Name Service (DNS). Some of these protocols are used directly by users as applications, such as FTP and Telnet. Other protocols are used directly by the applications, such as SMTP and HTTP. Other protocols, such as DNS, are used indirectly or are used by the programs and operating system routines.

Most of the application layer protocols are simple text conversations that use American Standard Code for Information Interchange (ASCII) code. This coding was adopted because it is a general code that almost any type of computer can understand. Thus, the protocol data can pass through almost any type of hosts, gateways, and routers. The downside of using text conversations is that eavesdropping can give out all the information about the communication session. The eavesdropped data can be easily comprehended and that is not an acceptable thing in many cases. To cover that, secure versions of these protocols were created. Secure HTTP (HTTPS) and Secure Socket Layer (SSL) are used to verify the identities of the hosts at both ends of a HTTP conversation and to encrypt data in transit between them. This is only an example of how the security issues were handled and there is a broad range of solutions that are currently in use.

3.6.1 Hyper Text Transfer Protocol

HTTP is a protocol used to exchange text, pictures, sounds, and other multimedia data using a Graphical User Interface (GUI). HTTP is the building block of the World Wide Web (WWW). It was adopted in 1990 in the Internet. It was one of the reasons of the wide spread of the Internet around the world [10].

In RFC 1946 that identified HTTP v1.0, HTTP was defined as "an application-level protocol with the lightness and speed necessary for distributed, collaborative, hypermedia information systems". HTTP is used in the WWW to transfer *web pages*. These pages are usually written with a format language called Hyper-Text Markup Language (HTML). As stated earlier, these pages can contain text, pictures, audio, video, and other multimedia formats. On the user end software named *Web Browser* is used. The web browser is a client-software used to view web pages via GUI. The main purpose of the browser is to translate (or render) the HTML pages into easy user-comprehendible graphical content.

HTTP is considered one of the most commonly used request-response (or sometimes called client–server) protocols. The web browser initiates a request and creates a TCP session with the web server that contains the web page. During this session the web page is transferred to the client host and it is translated to the GUI of the web browser to be seen by the user.

HTTP uses TCP as its transport protocol because TCP is reliable, unlike UDP.

The port number used by HTTP is port 80. The exact page or resource to be accessed on the web server is identified by a *Uniform Resource Identifier* (URI). The HTTP exact resource addresses are usually called *Universal Resource Locator* (URL). Typically the URL starts with a (http:) for example http://www.springer. com.

Each webpage consists of a main container, which is the main HTML page, along with other items within the page called *Referenced Objects*. Each object has a unique URL that identifies its exact location and name on the Internet.

Generally, there are two modes of operation by which HTTP fetches the page and all of its referenced objects; persistent and non-persistent HTTP.

In *persistent* HTTP, a single TCP connection is used to fetch the page container and all of its referenced objects. The TCP connection is opened to fetch the main page container and then it is left open and the client starts sending requests to fetch the referenced objects one after the other. The up side of this mode is that it does not cause much load on the server while the down side is that it can be slow if there is a lot of objects as it is possible to fetch only one object at a time.

In *non-persistent* HTTP, a separate TCP connection is created for each object. At first a TCP connection is setup to fetch the main HTML container. Then, the connection is closed. The client reads the main container and identifies the referenced objects and starts creating a new TCP connection for each object separately. Logically, this process looks slower than persistent HTTP because of the time required to create a new TCP connection for each object. However, the process can be multiple times faster if the server is setup to accept more than one

Table 3.2 Request messages used in HTTP/1.0 and HTTP/1.1

Version	Message	Function
HTTP/ 1.0	GET	Requests to fetch a web page or a referenced object from the server to the client
	POST	Sends form information from client to server. For example, when a user needs to login to a website, a username and password needs to be passed to the server
	HEAD	Asks server to leave requested object out of response
HTTP/ 1.1	GET	Requests to fetch a web page or a referenced object from the server to the client
	POST	Sends form information from client to server. For example, when a user needs to login to a website, a username and password needs to be passed to the server
	HEAD	Asks server to leave requested object out of response
	PUT	Uploads a complete file to a specified location in the server. For example, uploading an attachment to your web-based email service
	DELETE	Deletes a file located in the server. For example, deleting a temporary session file stored on the server when a client wants to log out of the website

TCP connection from the same client simultaneously. If the client requests multiple objects at the same time, the whole process would be much faster than persistent HTTP. The down side of this mode is that more load would be on the server because it will have to handle higher number of simultaneous TCP connections. This can degrade performance severely if the server is a busy one.

The HTTP version that is currently used is called HTTP/1.1 that was defined by RFC 2068 [11] in 1997 and later evolved to RFC 2616 in 1999 [12]. The major upgrades in HTTP/1.1 from HTTP/1.0 include the way HTTP handles caching; how it optimizes bandwidth and network connections usage, manages error notifications; how it transmits messages over the network; how internet addresses are conserved; and how it maintains security and integrity. There are also differences in the methods, or messages, used.

HTTP uses message formats that are human readable and can be easily understood. Table 3.2 shows the differences between request messages used in HTTP/1.0 and HTTP/1.1.

Table 3.2 shows that uploading a file to a web server was not possible before HTTP/1.1. We can say the same about deleting a file on the server. These two methods that were introduced in HTTP/1.1 added valuable functionalities to the protocol that became essential in our current daily life.

In situations where the amount of data a client needs to pass to the server is minimal, this data is encoded in the URL instead of invoking PUT method. For example, when a client searches something on Google, the search query is passed in the URL like http://www.google.com/search?q=osi when the client is searching for "osi".

Table 3.3 Commonly used responses of HTTP server

Response code	Meaning
200 OK	Successful request. The requested object is sent with this message
301 Moved permanently	The requested object has moved to a different location. The new location is identified in the same message
400 Bad request	Request messages is formatted in a wrong way and the server does not understand it
404 Not found	Requested object was not found. Most browsers translate this response to "This page cannot be displayed" or "Page not found"
505 HTTP version not supported	Requested HTTP version is not supported by this server

Server responses can vary depending on the request type and the availability of resources. Table 3.3 shows the most commonly used server responses in HTTP.

3.6.2 Domain Name Service

Accessing services using IP addresses can be troublesome to users. The user will have to remember the IP address of all the websites and email services he or she would like to use. DNS was designed to associate an easy-to-remember address to the IP address of the server. For example, this way you will not have to remember 173.194.36.33, but instead you will remember www.google.com.

DNS was identified in RFC 1034 and RFC 1035 [13, 14].

DNS works behind the curtains for almost all other application layer protocols, such as HTTP, FTP, Telnet, etc. What actually happens when a user writes a URL in the web browser is that the browser sends a DNS request to the DNS server that contains the domain name. The DNS server replies with the IP address of the web server that hosts the domain's website. Then, the browser communicates with the web server and views the web page. The meanings of a sample URL parts are shown in Fig. 3.10.

The first part of the URL defines which protocol is used to access this resource. This can be HTTP, HTTPS, FTP, etc. The second part of the URL defines the exact host that contains this resource. Most web sites have (www.) in this part. In some web sites, this part is called a *subdomain*. The subdomain is an integral part of the site that is devoted to a certain part of the site. The site owners goal is to make parts of their websites easily accessible to users without having to visit the central website represented in the (www.) part of the domain or the domain root. The third and the fourth parts of the URL represent the complete domain name. This part of the name is chosen by the site owner. The site owner registers the domain name with a licensed *domain name registrar*. The root DNS servers are then updated so that each time a user writes the domain name in the browser, the website associated with it is shown.

http://	www.	springer.	com	/computer/
protocol type	destination host	domain name	TLD type	destination folder

Fig. 3.10 The parts of a sample URL

The fourth part of the URL is usually called *Top Level Domain* (TLD). Some TLDs represent a generic name or a country code. The generic name usually represents the type or the purpose of the website. The most common generic names are:

1. .com: For commercial websites.
2. .net: It was intended for the use of network service providers.
3. .org: It was intended for all organizations not fitting in other categories.
4. .edu: For educational institutes and websites.
5. .gov: For governmental institutes.

Until the end of the year 2013, there were more than 350 TLDs available on the Internet according to *Internet Assigned Numbers Authority* (IANA), which is the organization responsible for registering the TLDs in the world. Most of these TLDs represent country codes. For example, .uk is the country TLD for the United Kingdom. Recently, IANA have opened the door to registering TLDs for different purposes like .inc, .music, .news, .blog, etc.

DNS was designed to operate as a hierarchical system. It was not designed to be a centralized service for the following reasons;

1. A centralized DNS service would mean that there will be a single point of failure. Since almost all other application layer protocols rely on DNS, the failure of the centralized DNS would mean the failure of the Internet.
2. Traffic volume to this centralized DNS would be huge.
3. Maintenance and scaling would be a genuine problem that jeopardizes the continuity of the service.
4. Choosing any location around the world to place this central DNS server would leave it far from other parts of the world. This means that the response time, and hence the service quality, would deteriorate the farther you get from the centralized server.

Due to the aforementioned reasons, DNS was chosen to be decentralized into a hierarchical design. Figure 3.11 shows the hierarchy of the DNS service.

In Fig. 3.11 you can see that the servers at the lowest level of the DNS hierarchy are called Authoritative Servers. An *Authoritative Server* is a DNS server responsible for maintaining and providing DNS records of one or more domains. When an organization have more than one server, such as a web server and a mail server, an authoritative server is required. The authoritative server keeps track of the IP addresses of each server in the organization's network such that when a client (from outside or inside the network) needs to communicate with any of the

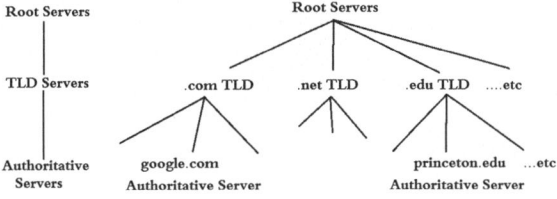

Fig. 3.11 DNS hierarchy

organization's servers, the client gets the IP address of this server from the authoritative server. For example, when a client needs to connect to scholar.google.com, Google's authoritative server is the one to contact to get the IP address of scholar.google.com. Authoritative servers are usually maintained by the organization itself or sometimes it is outsourced to an Internet Service Provider (ISP).

TLD Servers are responsible for maintaining and providing information about Authoritative servers registered with a specific TLD extension. For example, .com TLD server is responsible for providing information about all Authoritative Servers managing domains that have a .com TLD. Country TLD servers are the ones responsible for managing the country's TLD like .uk, .us, .es, etc. These country TLD servers are usually managed by the telecommunication authority of this country. Generic TLDs are managed by large information technology organizations. For example, .com and .net TLDs are managed by VeriSign Global Registry Services [15].

Root Servers are the servers responsible for maintaining and providing information about TLDs and TLD servers. The number of these servers, until the end of year 2013, was 386 instances distributed around the world. Their location can be found in [16].

Another type of servers lay outside of the hierarchy is called Local DNS Servers. A Local DNS Server is the first DNS server a client contacts to ask for DNS record of a certain domain. The local DNS server's IP address is the one that you find in your host identified as the "Primary DNS Server" and "Secondary DNS Server". Usually, each ISP manages its own local DNS server. The local DNS server keeps copies of the TLD servers information cached in to send the information to the clients directly.

DNS query is executed in one of two methods; Iterated-query, and Recursive-query. In *Iterated-query*, the request is forwarded as follows:

1. The client forwards the query to the local DNS server.
2. The local DNS server forwards the query to one of the root servers.
3. The root server replies "I don't know, but ask this server", and attaches the IP address of the TLD server responsible for the domain in question.
4. The local DNS server forwards the query to the TLD server address received from the root server.
5. The TLD server replies "I don't know, but ask this server", and attaches the IP address of the authoritative server responsible of the domain in question.

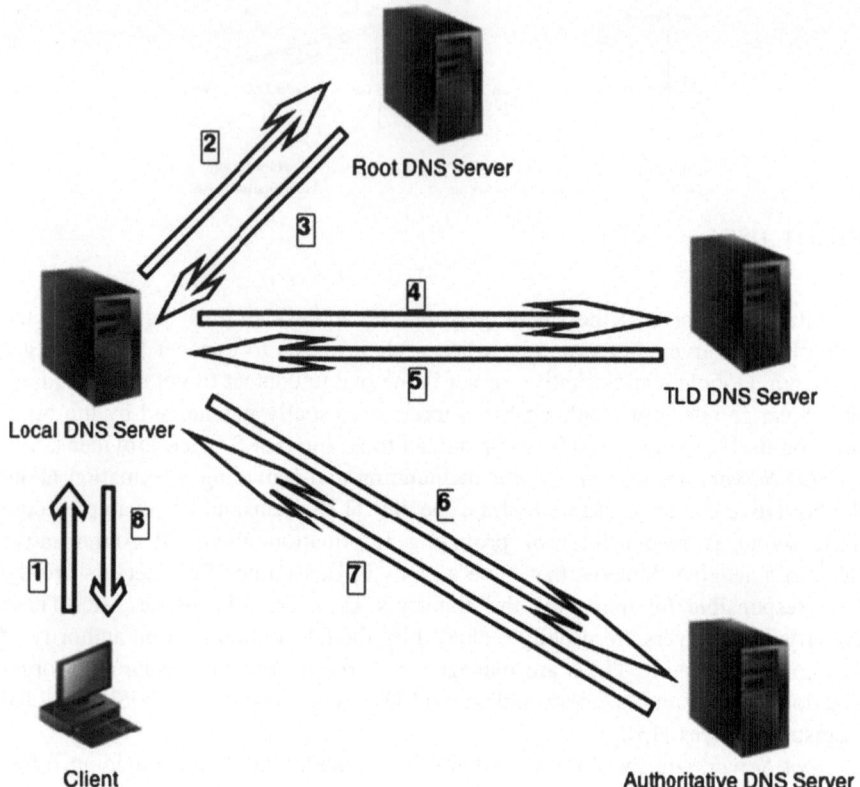

Fig. 3.12 DNS iterated-query

6. The local DNS forwards the query to the authoritative server that is responsible for the domain in question.
7. The authoritative server replies with the DNS record to the local DNS server.
8. The local DNS server forwards the DNS record received from the authoritative server back to the client. The local DNS server also caches the record in its DNS cache to forward it to other clients when they request it.

Figure 3.12 shows the DNS query flow in iterated-query.

Looking at Fig. 3.12, it can be seen that most of the work done to retrieve the DNS record was done by the local DNS server. If the local DNS server is serving a large number of users, out of 2.4 billion Internet users according to [17], its performance can degrade and responses would not be delivered on time. On the other hand, root, TLD, and authoritative servers handle only a small fraction of the work.

Recursive-query operates a follows:

1. The client forwards the query to the local DNS server.
2. The local DNS server forwards the query to one of the root servers.

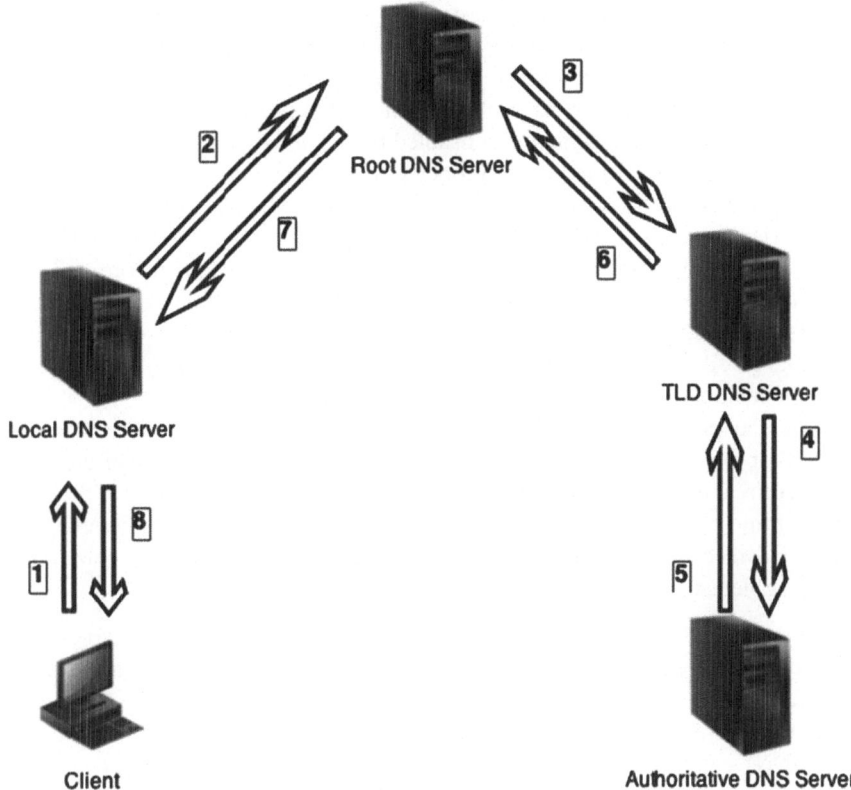

Fig. 3.13 DNS recursive-query

3. The root server keeps the session open with the local DNS server and contacts the TLD server responsible for the domain's TLD through a new session.
4. The TLD server keeps the session open with the root server and contacts the authoritative server responsible for the domain through a new session.
5. The authoritative server responds to the TLD server's request with the requested DNS record.
6. The TLD server forwards the DNS record to the root server requesting it.
7. The root server forwards the record to the local DNS server that requested it.
8. The local DNS server forwards the DNS record to the client.

Figure 3.13 shows the flow of the DNS query in recursive-query DNS.

Looking at Fig. 3.13, it can be easily seen that the server handling most of the process in recursive-queries is the root server. Root servers will have to handle two sessions for each request since it is received, until the record comes back from the authoritative server through the TLD server. All requests of all domains would have to pass through the root servers, while only part of the requests go to the TLD server (because it handles only requests for 1 TLD), and an even smaller portion will go to the authoritative server (because it handles requests for a specific domain only, or a few domains). This also implies that servers operating in recursive mode will need to have more memory and processing power.

Recursive-queries are sometimes disabled by the network administrators due to security issues. If the server is not properly secured, malicious attackers can use the compromised DNS server to flood other DNS servers as part of a Denial of Service (DoS) attack.

The positive side of this method of operation in DNS is that the record will pass through all the servers on the way to the client. Hence, it will be cached in all of these servers such that the next request for the same domain will be served without having to go all the way to the authoritative server. This makes the query response more efficient and reduces the load of authoritative servers.

From the previous discussion we conclude that it is not possible to select iterated- or recursive-query as the better mode of operation. Each one of them has its own pros and cons and can perform better in certain scenarios. Most of the time, you will find a mixture of both used in different levels of the DNS hierarchy.

A DNS record, sometimes referred to as Resource Record (RR), contains the following fields:

1. Type: A number identifying the type of the record. Records can carry information about email exchanges, name servers, ...etc. In total, there are over 35 different types of DNS records.
2. Name: Name of the host. This name is a Fully-Qualified Domain Name (FQDN).
3. Class: It is set to IN (short for Internet) for regular DNS records involving Internet hostnames, servers, or IP addresses. Other classes also exist such as Chaos (CH) and Hesiod (HS).
4. RLength: Length of the RData field. Used because RData can have variable length.
5. RData: The record data that needs to be passed to the client, such as IP address for address records, or priority and hostname for mail exchange records.
6. TTL: Short for Time-to-live. This field indicates the lifetime of the record. This is particularly important for record caching. When the record life time expires, the record is deleted and a new "fresh" version is requested when needed.

As stated above, there are over 35 different types of DNS records or RRs. Table 3.4 shows the most commonly used record types.

It is clear that DNS is not only about mapping domain names to IP addresses. Aliasing, identification of different services in servers, along with many other benefits of DNS exist. DNS can also be used for load distribution. This is best explained through an example;

Maintaining high performance in servers that provide services to large number of clients has always been a challenge. Some services are too large to be handled by a single server such as the webpage of www.google.com. This requires having

Table 3.4 Most common DNS record types

Record type	Usage
A	Used to identify the IP address of a host. This host can be a web server or any other server
MX	Used to identify the IP address of the Mail eXchange server
CNAME	Used to identify Canonical NAME. This type of records allows the use of alias names for canonical domain names. For example, when you use www.google.com the actual FQDN of the server can be googleweb.eastcoast.dc1.googlehosted.com
NS	Name Server record is used to identify the IP address of the authoritative DNS server responsible for maintaining DNS records of this domain

more than one server mapped into the same domain name. The load can be distributed (somehow evenly) using the authoritative DNS server of Google. When millions of requests come to the authoritative server asking for the IP address of the host www that is connected to the domain google.com, the server supplies an IP address of a different server each time. When each request forwards the IP address of a different server from the server farm that is hosting the website, the load will be distributed over these multiple servers and no single server would be serving all of these requests. This method is not bullet-proof, of course. DNS caching in root, TLD, and local DNS servers will cause uneven distribution of load because they will supply the IP address of the same web server to all the requests arriving at this root, TLD, or local DNS server.

3.6.3 File Transfer Protocol and Trivial File Transfer Protocol

FTP is an application layer protocol specialized in transferring files from one host to another. It relies on TCP at the transport layer. Thus, it is a reliable data transfer protocol. This protocol can be used in two ways; directly by user and indirectly by programs. The user can use FTP directly by typing its commands or by using a browser. Also, some programs, namely FTP Clients, use FTP to copy files from one host on a network to another.

According to the current FTP standard [18], FTP has gone through many evolution steps starting from IETF RFC 114 in 1971 and ending in the current IETF RFC 959.

FTP uses two separate connections for control and data transfer. At the user login, the client host connects to the FTP server on port 21 to setup the control connection. After the setup of the control connection, FTP initiates another TCP connection to transfer the data using port 20. When the data transfer is over, the data connection is terminated automatically, but the control connection remains

Table 3.5 Commonly used FTP commands

Command	Use
USER	Sends the username to the FTP server for authentication before the connection is setup
PASS	Sends the password after the username to the FTP server to complete the authentication phase
LIST	Request to display the files and folders contain the current active location. FTP servers are capable of remembering the current active location because FTP is a stateful protocol
RETR	Download a file from the server to the client
STOR	Upload a file from the client to the server

Table 3.6 Commonly used FTP server responses

Response code	Meaning of code
331	Username is OK, send the password to complete the authentication phase
125	Data connection already open; transfer starting
425	Data connection cannot be opened. This response is to indicate a problem with opening a new data transport session on port 20
452	Error writing file. This response message means that the file that the client is trying to upload was not written properly. This can be due to an error in transmission or there is no permission to write in the current folder

active in case of another data transfer. The control connection remains active until the user ends it by logging off.

Just like HTTP, FTP also has response codes to respond to the commands sent by the client. FTP messages, like HTTP's, are human-readable as shown in Table 3.5. Table 3.6 shows the most common FTP server responses.

Unlike HTTP, FTP is a stateful protocol where the server maintains the user state. FTP provides advanced features like the ability to browse the server folders, the ability to create, rename, and delete folders at the server side, the ability to delete, and rename files on the server, and the ability to change properties of files and folders on the server.

FTP uses two modes of file transfer; ASCII and binary. The mode of transfer determines the type of encoding used when transferring the files.

Despite the fact that a user can use FTP protocol directly from a browser, usually users use a FTP client program. This program provides the user with more controllability over the communication session.

FTP provides the ability to protect the connection with a username and a password. This way, only authorized users can create communication sessions with servers. Most FTP servers can provide the ability to assign folders to users such that each user sees only the folder he or she is allowed to.

FTP has been a target for different types of security attacks for a long time. Thus, it is currently considered insecure to use the plain old FTP [19].

Many standards and protocols were developed to address the security weaknesses found in FTP. This includes IETF RFC 2228 [20], IETF RFC 4217 [21], and SSH FTP [22] that was introduced as a draft and never became a standard.

On the other hand, there is a simplified version of the FTP protocol called *Trivial File Transfer Protocol* (TFTP). TFTP, unlike FTP, is an unreliable protocol because it relies on UDP as its transport layer protocol. The only form of reliability found in the TFTP is that each non-terminal packet is acknowledged separately. According to [23], TFTP was designed to be small and easy to implement. Thus, TFTP lacks a lot of the features that FTP provides. It can only read and write files to and from a remote server. TFTP uses the UDP port number 69.

TFTP uses a different connection mechanism than FTP's. TFTP does not have two separate connections for control and data. Any transfer begins with a request to read or write a file, which also serves as a connection request. If the server grants the request, the connection is opened and the file is sent in blocks of fixed length of 512 B. Each data packet contains one block of data, and must be acknowledged by an acknowledgment packet before the next packet can be sent. A data packet of less than 512 B marks the end of the transfer and signals termination of the connection.

The simplicity of TFTP made it easily implementable in hardware. Many networking devices, such as routers, adopted TFTP as part of their hardware to copy setting and operating system files. This is due to the fact that FTP hardware implementation would be complex and has many features that are not needed in these particular cases of hardware implementation.

Taking into consideration that TFTP is not as reliable as FTP, TFTP is better to be used in local networks that are stable with low error rates. Although TFTP can be used on the Internet, it is still better to use FTP when transferring data over internetwork systems.

3.6.4 Simple Mail Transfer Protocol

SMTP is a reliable application protocol used to transfer *electronic mail* (email) from a client to email server and between email servers. It was first adopted as a standard in 1982 in RFC 821 [24] and later developed into RFC 2821 in 2001 [25].

The email is transferred from the client software to the email server using a reliable TCP session made especially for the SMTP protocol. If the destination email is on the same server, the email is stored in the mailbox of the destination user. If the destination email is on another email server, the email is relayed to the destination server using SMTP as well. The email is then stored in the mailbox of the destination user waiting to be picked up.

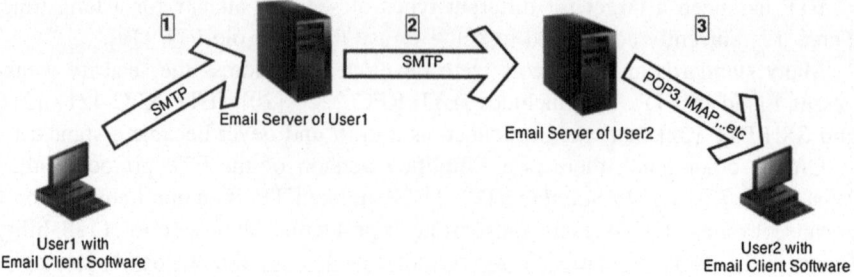

Fig. 3.14 Protocols used in sending and receiving electronic mail

There are many protocols that retrieve emails from the users' mailboxes and deliver them to the user's client software. The most known protocols for email access and retrieval are Post Office Protocol v3 (POP3) and Internet Message Access Protocol (IMAP). Figure 3.14 shows the protocols used in an email transfer from one user to another user.

In Fig. 3.14, the Email Server of User1 reads the destination email address to check whether it is a local email within the same server or it is hosted by another server. If the destination email addresses is a local address, i.e., within the same domain, it is stored directly into the destination account's mailbox. If the destination email address is in a different domain, it is forwarded to the SMTP queue of outgoing messages waiting to be sent over to Email Server of User2.

Each email server has a group of mailboxes that are used to store emails sent to users who have accounts in this server. The size of the mailbox as well as the email retrieval protocols allowed are decided by the settings of server. Other settings such as the maximum message size, supported encoding types, allowed attachment file types, ...etc. also exist in the server.

The old SMTP protocol did not offer a satisfying amount of security. It did not require authentication neither. The administrators always needed to take some actions to provide security to their email services. For example, the administrator can prevent hosts that are not part of the local network to use the SMTP server to send or relay mail. This action will prevent unauthorized users from using the SMTP servers as mail relays.

Most of the SMTP servers nowadays use a username-password authentication mechanism in sending the emails. Each user must enter a correct username and password to be authorized to send emails. Most servers also now support encryption protocols like TLS to assure security of the sent messages.

SMTP transfers emails in ASCII format and uses the TCP port number 25. Despite the fact that email is transferred in ASCII format, the user can send any type of files attached to the email; such as audio, video, photos, ...etc.

SMTP protocol gives the user the ability to send the email to more than one recipient. The email sender can send carbon copies and blind carbon copies of the email to other recipients.

Table 3.7 Most commonly used SMTP messages

Message	Usage of message
Connect	The first message sent from the client to the server asking the server to initiate a connection to send an email
HELO	Greeting the server to see if it is possible to send a mail message
MAIL	Stating the mail address of the sender
RCPT	Stating the mail address of the recipient. This message can be repeated more than once if the mail is intended for more than one recipient
DATA	The contents of the mail message

Table 3.8 Most commonly used SMTP server responses

Response	Usage of response
220	Service is ready for the domain. This code is sent as a response to the connect message
250	The requested mail action is accepted. This code is sent in response to the HELO, MAIL, RCPT and other messages to indicate the mail transfer process is going smoothly
354	Start mail input and end with a ".". The server will keep receiving parts of the mail coming through DATA messages until a message is received with a "." In the end. This "." indicates the end of the mail message
421	Service not available, closing the transmission session. This code is sent in response to refusal of connect, HELO, MAIL, RCPT, or DATA messages

The SMTP version of 2001 is sometimes known as Extended SMTP, or ES-MTP. One of the extensions added to the protocol was the Delivery Status Notification (DSN). This extension notifies the sender when the email could not delivered and what was the reason preventing the delivery.

SMTP also uses human-readable messages. The SMTP standard contains over 15 request messages. Table 3.7 shows the most commonly used commands from the client to the SMTP server and Table 3.8 shows the common SMTP server responses.

Other response messages are used to point out that something has went wrong in the transmission session.

Electronic mail messages remain in the mailbox of the destination user until it is accessed or retrieved. As stated earlier, many protocols can be used to access the mailbox and retrieve emails such as POP3, and IMAP. In some cases, such as web mail services like Gmail, Yahoo mail, …etc., HTTP can be used.

What actually happens is that HTTP provides only a user interface to the mailbox while the actual email operations, like deleting, marking as read, moving, …etc., are done in the background by IMAP.

IMAP is much more complex protocol as compared to POP3. IMAP gives the ability of mailbox synchronization among different mail client software. This means that your mailbox looks the same using any mail client software, unlike POP3. In POP3, emails are downloaded by the mail client software such that if you access the mailbox from another client computer, you would find it empty.

However, the details of the operation of IMAP are beyond the scope of this brief.

3.6.5 Post Office Protocol Version 3

POP3 protocol is a reliable application layer protocol that is used to retrieve emails from the email server to the client. The mail transfer mechanism mentioned in the SMTP subsection is shared between POP3 and SMTP. SMTP is responsible for forwarding the email from the sender to the SMTP server, and between SMTP servers. When reaching the destination email server, if the destination is not already in the same server, POP3 protocol takes over. POP3 protocol is the one responsible for retrieving the emails by users.

As stated in [26], POP3 is not intended to provide manipulation operations of email on the server. Usually, the email message is downloaded to the user's computer by the mail client software and then deleted from the server. Some POP3 servers provide extra features like keeping the message in the server for few days after being downloaded.

POP3 relies on TCP as its transport protocol. Thus, it is a reliable protocol. POP3 operates in port number 110. When the POP3 service is enabled, the POP3 servers keep listening to port 110. When a user wishes to use the services of the server, for example to receive emails, the user initiates a conversation with the server. Then, the user starts giving commands to the server and waits for responses. As in many application protocols, the POP3 protocol commands are ASCII-encoded text keywords followed by one or more arguments.

Usually, POP3 servers use a username-password authentication scheme. Most servers use the mailbox name as the username. Some of the POP3 servers nowadays impose high complexity in user passwords to prevent users' mailboxes from being hacked. When POP3 was first used, it used to transfer the passwords unencrypted. Currently, most POP3 servers use some sort encryption mechanism to transfer passwords, such as one-way hash functions.

Tables 3.9 and 3.10 show the most common POP3 messages and server responses, respectively.

3.6.6 Telnet

Telnet, previously called *Network Terminal Protocol*, is a reliable application layer protocol that uses TCP to establish a bi-directional ASCII-encoded text conversation [27]. This conversation is usually used to do a virtual terminal process across a network or Internet. In [27], Telnet's purpose is clearly stated as, to provide a fairly general, bi-directional, byte oriented communications facility. Its primary goal is to allow a standard method of interfacing terminal devices and terminal-oriented processes to each other. It can be used for host-to-host communication and process-to-process communication (distributed computation).

For a long period of time Telnet was, and still, an essential part of networks and applications. Telnet was used before to send mail, transfer files, as well as virtual terminal applications. Until now, Telnet is a very useful tool for network

Table 3.9 Common POP3 messages

Messages	Usage of message
USER	Send the username
PASS	Send the password in plaintext. Another command "APOP" is used to send the password in hashed format for security purposes
LIST	Requesting a list of email messages available in the mailbox
RETR	Download a particular email message
DELE	Delete a particular email message from the mailbox. This message is usually used after downloading the email message

Table 3.10 Common POP3 server responses

Response	Meaning of response
+OK	This response code indicates that the requested operation was done successfully. This response is sent after successfully finishing any of the operations requested by any message type
−ERR	This response code indicates that the requested operation failed. This response is sent along with the details after failing to complete a request sent from the client

administrators and users in many aspects. Telnet protocol can be used to configure and manage networking devices remotely, such as routers. It can be used in troubleshooting many types of servers, such as email servers. It can be used as a way of checking all-layers connectivity between networking hosts and/or devices.

As most of the application protocols do, Telnet uses simple ASCII-encoded text commands. And most of the servers that allow users to Telnet them from the Internet impose a username-password authentication scheme.

Telnet protocol as stated earlier, relies on TCP as its transport protocol and uses TCP port number 23. Most of the advanced Telnet client software allows the user to choose a port number other than 23. This feature has great deal of importance in security. The server and the authorized users can agree to use a different port number for the Telnet applications. Then, an intruder will not be able to access the server easily even if he or she had hacked a username and a password.

Since its earliest version in the 1970s, the Telnet standard was extended with over 20 different RFC discussing different aspects of this important protocol.

Security has always been a concern when using Telnet. The usernames and passwords are passed through in plaintext without any encryption processes. Thus, Telnet was thought of as an insecure protocol and the focus has shifted to a more advanced protocol named Secure Socket Shell (SSH). SSH was identified in RFC 4251 [28]. SSH creates an encrypted tunnel between the two communicating sides and is generally used as a more secure alternative to Telnet, especially when the connection is done over the Internet. Telnet can be used in more secure environments within local area networks to perform testing tasks. For example, Telnet can be used to test connectivity, or even to test email servers as in [29]. It can also be used to configure networking devices like routers and switches [30].

References

1. Cerf, V.G., Kahn, R.E.: A protocol for packet network intercommunication. IEEE Trans. Commun. **22**(5), 637–648 (1974)
2. Russell, A.L.: OSI: The Internet That Wasn't [Online]. http://spectrum.ieee.org/computing/networks/osi-the-internet-that-wasnt (2013)
3. IETF: RFC 791, California, Standard (1981)
4. IETF: RFC 792, California, Standard (1981)
5. IETF: RFC 826, California, Standard (1982)
6. IETF: RFC 903, California, Standard (1984)
7. IETF: RFC 2390, California (1998)
8. IETF: RFC 793, California (1981)
9. IETF: RFC 768, California, Standard (1980)
10. IETF: RFC 1945, California (1996)
11. IETF: RFC 2068, California (1997)
12. IETF: RFC 2616, California (1999)
13. IETF: RFC 1034, IETF, California, Standard (1987)
14. IETF: RFC 1035, IETF, California, Standard (1987)
15. IANA: Root Zone Database [Online]. http://www.iana.org/domains/root/db (2013)
16. IANA: Root Servers [Online]. http://www.root-servers.org/ (2013)
17. Internet-World-Stats: World Internet Users Statistics [Online]. http://www.internetworldstats.com/stats.htm (2012)
18. IETF: RFC 959, California, Standard (1985)
19. Harris, B., Hunt, R.: TCP/IP security threats and attack methods. Comput. Commun. **22**(10), 885–897 (1999)
20. IETF: RFC 2228, California, Standard (1997)
21. IETF: RFC 4217, California, Standard (2005)
22. IETF: SSH File Transfer Protocol [Online]. http://tools.ietf.org/html/draft-ietf-secsh-filexfer-02 (2001)
23. IETF: RFC 1350, California, Standard (1992)
24. IETF: RFC 821, California, Standard (1982)
25. IETF: RFC 2821, California, Standard (2001)
26. IETF: RFC 1939, California, Standard (1996)
27. IETF: RFC 854, IETF, California, Standard (1983)
28. IETF: RFC 4251, IETF, California, Standard (2006)
29. Microsoft: How to use Telnet to test an Internet Mail Connection [Online]. http://support.microsoft.com/kb/196748 (2013, Dec)
30. Alani, M.: Guide to Cisco Routers Configuration: Becoming a Router Geek. Springer, London (2012)